International Conference on Aeronautical, Aerospace and Mechanical Engineering (AAME 2018)

MATEC Web of Conferences Volume 202 (2018)

Sarawak, Malaysia
28 – 30 June 2018

Editor:

Tzong-Hann Shieh

ISBN: 978-1-5108-7249-3

Printed from e-media with permission by:

Curran Associates, Inc.
57 Morehouse Lane
Red Hook, NY 12571

Some format issues inherent in the e-media version may also appear in this print version.

Printed by Curran Associates, Inc. (2018)

For additional information, please contact EDP Sciences – Web of Conferences
at the address below.

EDP Sciences – Web of Conferences
17, Avenue du Hoggar
Parc d'Activité de Courtabœuf
BP 112
F-91944 Les Ulis Cedex A
France

Phone: +33 (0) 1 69 18 75 75
Fax: +33 (0) 1 69 28 84 91

contact-edps@webofconferences.org

Additional copies of this publication are available from:

Curran Associates, Inc.
57 Morehouse Lane
Red Hook, NY 12571 USA
Phone: 845-758-0400
Fax: 845-758-2633
Email: curran@proceedings.com
Web: www.proceedings.com

TABLE OF CONTENTS

Preface

2018 International Conference on Aeronautical, Aerospace and Mechanical Engineering (AAME 2018) provided a forum for accessing the most up-to-date and authoritative knowledge from both industrial and academic worlds, sharing best practices in this exciting field. AAME 2018 was held in Curtin University, Malaysia, during the period June 28-30.

The event was held with presentations delivered by researchers and scholars from the international community, including keynote speeches and highly selective lectures.

The proceedings included papers which were from universities, research institutes and industries. All of the papers were subjected to peer-reviewing by conference committee members and international reviewers. The papers selected for the proceedings depended on their quality and their relevance to the conference. Studies presented in this volume cover the following topics: Mechanical Dynamics and Vibration, Mechanical Strength, Casting Technology and Equipment, Welding Technology and Equipment, The Machine Assembly Process, Non-metallic Processing, Aerospace Systems and Technology, Aerodynamics and Aeroelasticity, etc.

We would like to take this opportunity to thank many people. First and foremost we want to express our deep appreciation to keynote speakers, Prof. Ir. Dr. Mohd Hamdi Bin Abd Shukor from University of Malaya, Malaysia; Prof. Dr. Nawaf Hazim Saeid from Universiti Teknologi Brunei, Brunei; Prof. Sujan Debnath from Curtin University Malaysia, Malaysia; Prof. Andrew Ragai Henry Rigit from Universiti Malaysia Sarawak, Malaysia; Prof. Rosli Bin Ahmad from Universiti Tun Hussein Onn Malaysia, Malaysia.

We would like to thank the committee members for their hard work in making sure a smooth running of this conference. Many thanks to the reviewers for their excellent job to maintain the academic quality of this conference. Special thanks is expressed to Curtin University, Malaysia, who kindly gave us support.

Final thanks go to all authors and participants at AAME 2018 for helping to make it a successful event.

With our warmest regards,
AAME 2018 Organizing Committees
July 18, 2018

edp sciences

MATEC Web of Conferences

Statement of Peer review

In submitting conference proceedings to MATEC Web of Conferences, I certify to the Publisher that I adhere to the **Policy on Publishing Integrity** of the journal in order to safeguard good scientific practice in publishing.

1. All articles have been subjected to peer review administered by the proceedings editors.
2. Reviews have been conducted by expert referees, who have been requested to provide unbiased and constructive comments aimed, whenever possible, at improving the work.
3. Proceedings editors have taken all reasonable steps to ensure the quality of the materials they publish and their decision to accept or reject a paper for publication has been based only on the merits of the work and the relevance to the journal.

Title, date and place of the conference

2018 the International Conference on Aeronautical, Aerospace and Mechanical Engineering (AAME 2018)
Sarawak, Malaysia
June 28-30, 2018

Proceedings editor(s):

Prof. Vincent Lee Chieng Chen, Curtin University, Sarawak, Malaysia

Tzong-Hann Shieh

Date and editor's signature

12/9/18

17, avenue du Hoggar - PA de Courtabœuf – BP 112 - 91944 Les Ulis Cedex A (France)
Tél.: 33 (0)1 69 18 75 75 - Fax : 33(0)1 69 07 45 17 – www.edpsciences.org

Solidification, Microstructure and Mechanical Properties of Mg-2.8Nd-1.5Gd-0.5Zn-0.5Zr Cast Alloy with Erbium Addition

R. Ahmad[1], A.M.M. Elaswad[1], M. Z. Hamzah[1], N. R. Shahizan[1]

[1]*Department of Manufacturing and Industrial Engineering, Faculty of Mechanical and Manufacturing Engineering, Universiti Tun Hussein Onn Malaysia, Parit Raja, 86400, Johor, Malaysia*

Abstract. The thermal parameters of Mg-2.8Nd-1.5Gd-0.5Zn-0.5Zr cast alloy with 0.25 wt.% of erbium (Er) were evaluated by the computer-aided cooling curve thermal analysis(CA CCTA), whereas the microstructure analysis was investigated by the optical microscope and scanning electron microscopy. Results from the cooling curve and microstructure analysis showed that Er altered the grain size of the alloys. In addition, the grain size was reduced by approximately 19.6% with the addition of Er. Scanning electron microscopy results showed that Er formed an Mg-Zn-Nd-Er phase which distributed along the grain boundaries. Furthermore, the mechanical properties were investigated by hardness and tensile tests with Er addition. The addition of 0.25 wt.% of Er significantly improved the ultimate tensile strength and yield strength. In addition, the hardness value of Mg-2.8Nd-1.5Gd-0.5Zn-0.5Zr increased by 13.9% with Er addition.

1 Introduction

The request of weight reduction for structural materials has become increasingly strong in recent years[1]. Magnesium light alloys have aroused increasing interest in recent years for applications in aerospace, aeronautics and automotive [2],due to their high strength weight ratio and low density excellent castability and good machinability [3, 4].Rare-earth (RE) metals are added to Mg alloys to improve their strength, microstructure and creep resistance. Two groups of RE elements considered for Mg alloys are yttrium (Y), gadolinium (Gd), Erbium (Er) and dysprosium (Dy), which are highly soluble, and neodymium (Nd), cerium (Ce), and lanthanum (La), which are less soluble; both groups are used to improve the properties of Mg alloys [5]. Rare earth Er has a large solubility in Mg (6.56 at.%), whose beneficial effects in aluminium alloys has been recently identified [6]. In recent years, some novel magnesium alloys containing heavy rare earth elements such as Mg-Gd-Y-Zr, Mg-Dy-Gd-Nd, and Mg-Dy-Gd-Zr alloys have been developed [7]. It has been reported that the content variation of rare earth elements (RE) and element Er in the alloys can influence the phase formation and mechanical properties greatly [8]. The hard eutectic phases that formed as a result of rare earths addition also strengthen the alloy. Generally, magnesium strengthening through rare earth addition results from grain boundary strengthening caused by precipitates and solid solution. Mg–RE alloys are used in military aircrafts and aerospace applications [9]. Adding Erbium into a Mg–Zn–Zr alloy improves the deformability of the latter and leads to the formation of a fine and uniform microstructure. In addition, erbium can increase the thermal stability and improve the mechanical properties of the alloy structure [10]. This research aims to investigate the addition of Er separately on solidification □ehaviour, microstructure and mechanical properties of Mg-Nd-Gd-Zn-Zr cast alloy.

2 Experiment procedures

Magnesium cast alloy Mg-2.8Nd-1.5Gd-0.5Zn-0.5Zr has been used as a base alloy, 0.25 wt.% Erbium (Er) was separately added to the molten metal. The casting was carried out in an electric resistance furnace for a steel crucible under the protection of 2% SF6 and Ar [11, 12]. After the addition, the melt was stirred for approximately 5 min to ensure dissolution of the alloying elements [13]. The molten metal was then poured at 750 ° C ±10 into a preheated steel mould. The mould was then immediately covered to protect the metal from each reaction with air, while the temperature-time data was recorded during cooling to room temperature. The cooling-curve parameters (i.e., temperature and time) for the formation of the phases and the solidification behavior of the modified and unmodified alloys were determined by CA-CCTA. A K-type thermocouple was immersed from the top of the preheated stainless steel mold and placed at the center (Fig.1). The temperature–time data were recorded using a high-speed data acquisition system (EPAD-TH8-K) connected to a computer running the DEWE Soft 7.5 package at a dynamic rate of 100 Hz per channel. The smoothing and plotting of the first derivative of the cooling curves were carried out using the FlexPro 9.0 data analysis software for determining the phase solidification characteristics data. In order to ensure the reliability of the results, the thermal analysis was repeated three times for each composition. The samples for microstructure of the alloys were analysed using an Olympus optical microscope (OM) and scanning electron microscope (SEM). The grain size was evaluated using the linear intercept method, according to ASTM E112-10 and the values of the volume fraction evaluated by an image analysis method specified in ASTM E112-04. The standard of grinding and polishing were used on the specimens for OM observation and etched in (5 Ml acetic acid, 6g picric acid, 10 Ml distilled water and 100 Ml ethanol). The castings were machined and milled to obtain the precise specimen dimensions for the cylindrical tensile test according to ASTMB577M.

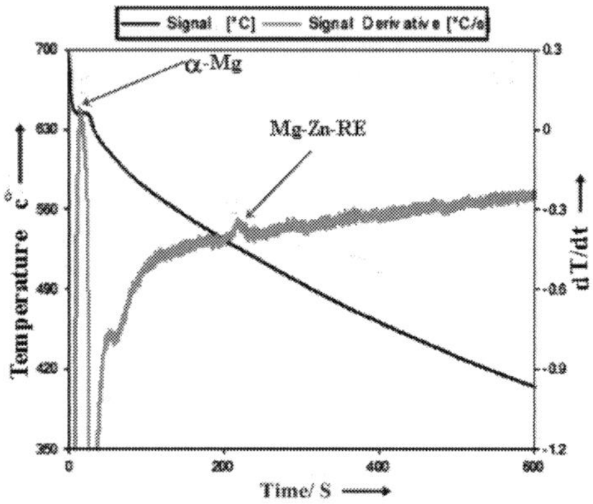

Figure 1. Illustration of mean solidification phases of Mg-2.8Nd-1.5Gd-0.5Zn-0.5Zr.

Table 1. Solidification characteristics of parameters determined during solidification of Mg-2.8Nd-1.5Gd-0.5Zn-0.5Zr alloy

Symbol	Description
$T_S^{\alpha\text{-Mg}}$	Start of α-Mg nucleation (liquidus) / °C
$T_{EU}^{\alpha\text{-Mg}}$	Eutectic growth temperature of dendrite α-Mg
$T_{Eu}^{Mg\text{-}Zn\text{-}RE}$	Formation of eutectic growth phase (Mg-Zn-RE) / °C
Te	End of solidification (solidus) / °C
ΔT	The total solidification temperature $\Delta T = (T_S^{\alpha\text{-Mg}} - Ts)$
Δt	The total solidification time $\Delta t = (t_S^{\alpha\text{-Mg}} - ts)$

3 Result and discussion

3.1 Cooling curves analysis

The obtained cooling curves were analyzed to identify the phase evolution of the base alloy systems during solidification with and without Er addition. The main cooling-curve parameters of the base alloy and Er containing alloys are listed in Table 2.

Table 2. Cooling-curve parameters of the base alloy and the base alloy with Er additions

Parameters	Alloys	
	0 wt.% Er	0.25 wt.% Er
$T_S^{\alpha\text{-Mg}}$ / °C	645.8	643.8
$t_S^{\alpha\text{-Mg}}$/s	11.01	10.01
$T_{EU}^{\alpha\text{-Mg}}$/ °C	635.9	627.7
$t_{EU}^{\alpha\text{-Mg}}$/s	29.18	25.34
$T_{EU}^{Mg\text{-}Zn\text{-}RE}$ / °C	526.8	526
$T_{EU}^{Mg\text{-}Zn\text{-}RE}$/s	218.7	201.9
$T_{S/°C}$	521.4	506.1
$t_{S/S}$	231.57	203.1

The solidification process of Mg alloys starts with the nucleation and growth of the primary α-Mg phase, followed by other phases containing the alloying elements. As shown Table 2, the α–Mg phase of base alloy was formed at 635.9 °C , and the solidification beginning at 645.8 °C ($T_S^{\alpha\text{-Mg}}$). The experimental results showed that when 0.25 wt.% Er is added, the nucleation temperature of α-Mg phase decreases, which obtained at 643.8 °C. It is reported that the addition of erbium into Mg–Zn alloys facilitated the formation of the α-Mg phase at around 638°C, and this phase appeared as the first peak, which corresponded to the solidification of the magnesium matrix [9, 14].

The eutectic growth temperatures of α-Mg phase $T_{EU}^{\alpha\text{-Mg}}$ decreased by addition of 0.25 wt. % of Er, which obtained at 627.7 °C. The decrease in the nucleation growth of α-Mg phase led to the decrease in the solidus temperature and simultaneously widened the solidification range [9].

Formation of the Mg-Zn-RE secondary phase is the last stage of solidification. The main characteristic parameters of the Mg-Zn-RE phase were determined using the first derivative and included the eutectic growth temperature $T_{EU}^{Mg\text{-}Zn\text{-}RE}$ and time $t_{EU}^{Mg\text{-}Zn\text{-}RE}$. Since the liquidus projection represents only the primary phase formed, it is necessary to know the primary and liquid phases in order to understand the solidification path, including which eutectic transformation takes place [15]. The results showed that the Mg-Zn-RE phase of base alloy was formed at 526.8 °C and With addition of 0.25 wt.% Er, the $T_{Eu}^{Mg\text{-}Zn\text{-}RE}$ temperature of Mg-Zn-RE phase decreased by 0.8 °C. The smaller magnitude of this effect compared with that of the previous addition may be due to the Er content in the intermetallic secondary phase [9].

The effect of Er addition on frozen time was obtained by, solidification time, $\Delta t = (t_S^{\alpha\text{-Mg}} - ts)$. The analysis results of the cooling curve parameters showed that the frozen time (solidification time) of the base alloy decreased from 220.56 s to 193.09 s with Er addition. This result is similar to that obtained by Fletcher, M., et al. [16], who reported that the decrease in liquid temperature was reflected on the grain size of the α-Mg phase. Er has an important effect on the solid solubility of

 https://doi.org/10.1051/matecconf/201820201001

Zn in the matrix, which changes as a result of the decrement of the solidus curve that shortens the nucleation time and reduces the grain size [8].

3.2 Microstructure analysis

The optical microstructures of the base alloy (a) and base alloy treated with Er (b) are shown in Figure 2. With Er, it greatly refines grain size of Mg-2.8Nd-1.5Gd-0.5Zn-0.5Zr alloy.

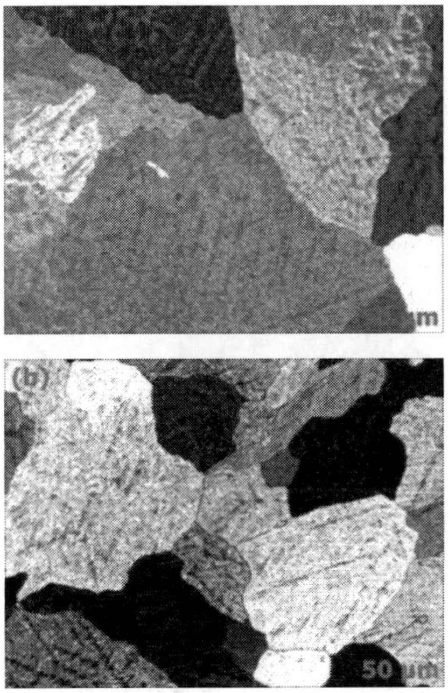

Figure 2. Microstructure images of (a) base alloy and (b) Er-modified alloy.

The addition of Er significantly decreased grain size from 316.8 μm to 254.4 μm. This grain refinement is due to the reduction of the liquid temperature and solidification time [17]. Similar observations were previously reported by Seetharaman, S., et al. [18], who reported that adding Er to the Mg–Al alloys significantly decreased the average grain size. Previous studies reported that the microstructures of Mg alloys could be refined by adding Er, which is a surface reactive element that restrains the growth of the secondary phase and thus reduces the solidification time [10, 14, 19]. These findings were confirmed by frozen time results in this investigation.

In addition, Er increased the volume fractions of the intermetallic secondary phase from 23.6% to 28.1%. The primary α-Mg grains were then enveloped by continuous networks of eutectic compounds, which have a strong tendency to detach when the grains are refined. Scanning electron microscopic (SEM) observations of base alloy treated with 0.25 wt.% of Er is shown in Figure 3. The SEM image of 0.25 wt.% Er shows the new intermetallic phase (Mg-Zn-Nd-Er) crystallized along the grain boundaries. These results can be attributed to the extremely low solubility of RE elements in the solid solution. The presence of these elements in Mg decreases the solubility of Zn in Mg. Thus, during solidification, the RE elements form the Mg–Zn–RE phase until the available RE is used without any formation of other compounds [20, 21].

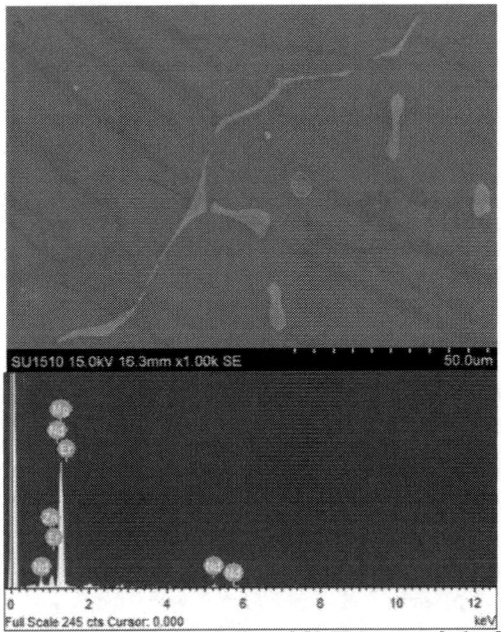

Figure 3 (a) SEM Micrographs and (b) EDS spot analysis of 0.25 Er containing alloy.

3.3 Mechanical properties

The effect of 0.25 wt.% Er on the tensile properties of the Mg-2.8Nd-1.5Gd-0.5Zn-0.5Zr alloy is shown in Table 3. The tensile strength had increase with 0.25 wt.% Er, which was improved by around 13.6%. Acceptable yield strength (YS) was also achieved, as shown in Table 3. The YS value reflects the alloy ductility, where it improved by 6 %. Moreover, the elongation index degraded 0.25 wt.% Er was added. The mechanical properties of the Mg alloys can be extensively influenced by alloying elements. Moreover, the strength of cast Mg alloys can be obtained through solid solution hardening, grain refinement, age hardening, and dispersion hardening [22, 23]. The solute atoms of Er in the Mg matrix can exert dragging forces on the grain boundary motion and decrease the recrystallization kinetics. The improved microstructure and mechanical properties can be attributed to the dissolved Er in the Mg matrix [14]. The tensile properties of Mg–Zn alloys are improved with the Er content. Er is highly soluble in Mg, which has a compact hexagonal structure. The solid solubility of Er in Mg decreases at decreasing temperatures. The atomic radius of Er is 29.2% larger than that of Mg, and thus Er addition leads to intensive solution strengthening in Mg. Increased solution strength and dispersion strength result in further increase in strength [8, 19].

Table 3. Tensile properties of base alloy treaded with Er.

Alloys	UTS, Mpa	YS, Mpa	EL %
0 wt.% Er	109	82.09	5.934
0.25 wt.% Er	123.91	87.04	3.843

The base alloy samples as cast and with 0.25 wt. % Er was examined to obtained the hardness value. However, it has indicated that the hardness value increases from 48.33 Hv for base alloy to 55.08 Hv for 0.25 wt.% Er respectively. The strengthening of Mg alloys is due to the solid solution of the borderline-sized RE atoms and second-phase hardening attributed to the Mg12RE intermetallic phase [24]. The mechanical properties of the Mg alloys are strongly influenced by their microstructure properties, such as grain size, volume fraction, and distribution of intermetallic phases [25].

4 Conclusions

The effect of the addition of 0.25 wt.% Er on the solidification characteristics, microstructure, and mechanical properties of Mg-2.8Nd-1.5Gd-0.5Zn-0.5Zr cast alloy was investigated. From the analysis, the following can be summarized:

1-The computer-aided cooling curve thermal analysis and microstructure results showed that Er improve solidification time and grain refinement.

2-Er decreased the solidification time of alloys by approximately 12% and thus decreased the grain size by 19.6%.

3-The mechanical properties of Mg-2.8Nd-1.5Gd-0.5Zn-0.5Zr had improved. The UTS improved by 13.6% and Ys improved by 6%. Moreover, the hardness value of the alloys increased with Er addition.

References

1. Zhang, Li, Zhang, Jinghuai, Leng, Zhe, Liu, Shujuan, Yang, Qiang, Wu, Ruizhi, and Zhang, Milin. "Microstructure and mechanical properties of high-performance Mg–Y–Er–Zn extruded alloy," Materials & Design (1980-2015). 256, **54**.(2014).
2. Wang, Qu Dong, Li, Da Quan, Li, Qiang, and Ding, Wen Jiang. "Effects of Er on Microstructure and Mechanical Properties of Mg-Zn-Er-Zr Magnesium Alloys," Materials Science Forum. 105, **546-549**.(2007).
3. Zhang, Jinyang, Xu, Min, Teng, Xinying, and Zuo, Min. "Effect of Gd addition on microstructure and corrosion behaviors of Mg–Zn–Y alloy," Journal of Magnesium and Alloys. 319, **4**.(2016).
4. Ma, Yutao, Zhang, Xingguo, Liu, Hongbin, and Meng, Linggang. "Effects of Er on the microstructure and properties of AZ31 magnesium alloy prepared via the EMS process," Rare Metals. 339, **29**.(2010).
5. Z. M. Sheggaf, R. Ahmad, M.B.A. Asmael, and A.M.M. Elaswad, and 1). "Solidification, microstructure, and mechanical properties of the as-cast ZRE1 magnesium alloy with different praseodymium contents," International Journal of Minerals, Metallurgy, and Materials**24**.(2017).
6. Rosalbino, F., Angelini, E., De Negri, S., Saccone, A., and Delfino, S. "Effect of Erbium addition on the corrosion behaviour of Mg–Al alloys," Intermetallics. 55, **13**.(2005).
7. Zhang, Jing, Zhang, Xu Feng, Pan, Fu Sheng, Liu, Chuan Pu, and He, Qu Bo. "Microstructure and Mechanical Properties of an Mg-Zn-Zr Alloy Modified by Rare Earth of Erbium," Materials Science Forum. 810, **610-613**.(2009).
8. Li, Han, Du, Wenbo, Li, Shubo, and Wang, Zhaohui. "Effect of Zn/Er weight ratio on phase formation and mechanical properties of as-cast Mg–Zn–Er alloys," Materials & Design. 259, **35**.(2012).
9. Ahmad, R, Sheggaf, ZM, and Asmael, MBA. "Effect of praseodymium and erbium additions on solidification characteristics, microstructure and mechanical properties of as–cast ZRE1 magnesium alloy," Materialwissenschaft und Werkstofftechnik. 218, **48**.(2017).
10. Zhang, Jing, Ma, Qi, and Pan, Fusheng. "Effects of trace Er addition on the microstructure and mechanical properties of Mg–Zn–Zr alloy," Materials & Design. 4043, **31**.(2010).
11. Pekguleryuz, Mihriban O, Kainer, Karl, and Kaya, A Arslan, 2013, Fundamentals of magnesium alloy metallurgy, Elsevier.
12. Katsarou, Lydia, Mounib, Maher, Lefebvre, Williams, Vorozhtsov, Sergey, Pavese, Matteo, Badini, Claudio, Molina-Aldareguia, Jon Mikel, Jimenez, Carmen Cepeda, Prado, Maria Teresa Pérez, and Dieringa, Hajo. "Microstructure, mechanical properties and creep of magnesium alloy Elektron21 reinforced with AlN nanoparticles by ultrasound-assisted stirring," Materials Science and Engineering: A. 84, **659**.(2016).
13. Sun, Ming, Hu, Xiaoyu, Peng, Liming, Fu, Penghuai, and Peng, Yinghong. "Effects of Sm on the grain refinement, microstructures and mechanical properties of AZ31 magnesium alloy," Materials Science and Engineering: A. 89, **620**.(2015).
14. Zhang, Jing, Li, Weiguo, Zhang, Baoxiang, and Dou, Yuchen. "Influence of Er addition and extrusion temperature on the microstructure and mechanical properties of a Mg–Zn–Zr magnesium alloy," Materials Science and Engineering: A. 4740, **528**.(2011).
15. Suzuki, A., Saddock, N. D., Jones, J. W., and Pollock, T. M. "Solidification paths and eutectic intermetallic phases in Mg–Al–Ca ternary alloys," Acta Materialia. 2823, **53**.(2005).
16. Fletcher, M, Bichlerl, L, Sediako, D, and Klassen, R, 2011, Compressive Creep Behaviour of Extruded Mg Alloys at 150° C.
17. Sheggaf, ZM, Ahmad, R, Asmael, MBA, and Elaswad, AMM. "Solidification, microstructure, and mechanical properties of the as-cast ZRE1 magnesium alloy with different praseodymium contents," International Journal of Minerals, Metallurgy, and Materials. 1306, **24**.(2017).
18. Seetharaman, Sankaranarayanan, Blawert, Carsten, Ng, Baoshu Milton, Wong, Wai Leong Eugene, Goh, Chwee Sim, Hort, Norbert, and Gupta, Manoj. "Effect of

erbium modification on the microstructure, mechanical and corrosion characteristics of binary Mg–Al alloys," Journal of Alloys and Compounds. 759, **648**.(2015).
19. ZHANG, Jing, HE, Qu-bo, PAN, Fu-sheng, ZHANG, Xu-feng, and LIU, Chuan-pu. "Effects of erbium on microstructure and mechanical properties of as-cast Mg-7Zn-3Al alloy," Transactions of Nonferrous Metals Society of China. s22, **18**.(2008).
20. Rzychoń, Tomasz, Szala, Janusz, and Kiełbus, Andrzej. "Microstructure, castability, microstructural stability and mechanical properties of ZRE1 magnesium alloy," Archives of Metallurgy and Materials. 245, **57**.(2012).
21. Rzychoń, T, Kiełbus, A, and Szala, J. "Microstructure and fluidity of sand cast ZRE1 alloy," Journal of Achievements in Materials and Manufacturing Engineering. 135, **26**.(2008).
22. Pekguleryuz, Mihriban O, Kainer, Karl, and Kaya, Aslan, 2013, Fundamentals of magnesium alloy metallurgy, Elsevier.
23. Friedrich, Horst E, and Mordike, Barry L, 2006, Magnesium technology, Springer.
24. Yang, Zhong, Li, JP, Zhang, JX, Lorimer, GW, and Robson, JAMSEL. "Review on research and development of magnesium alloys," Acta Metallurgica Sinica (English Letters). 313, **21**.(2008).
25. Muraliraja, R, Vettrivel, H, and Elansezhian, R. "Synthesis and Characterization of Magnesium Alloy added with Yttrium and To Study the Microstructure and Mechanical Properties," International Journal of Engineering and Innovative Technology. 338, **2**.(2013).

Mode I and Mode II Delamination of Flax/Epoxy Composite Laminate

Thamilarasu S. Rajendran[1], Mahzan Johar[1], Shukur Abu Hassan[1,2] and King Jye Wong[1]

[1]*Faculty of Mechanical Engineering, Universiti Teknologi Malaysia, 81310 Johor Bahru, Malaysia*
[2]*Centre for Advanced Composite Materials, Universiti Teknologi Malaysia, 81310 Johor Bahru, Malaysia*

Abstract. In recent decades, natural fibres are getting their attention as reinforcement in composite materials. This is because natural fibres are environmental friendly. However, delamination is commonly recognised as one of the earliest failures in composite laminates. The objective of the present work is to investigate mode I and mode II delamination behaviour of flax fabrics reinforced epoxy composite. The delamination characterisation was carried out using double cantilever beam (DCB) and three point end notched flexure (ENF) tests. The fracture toughness were calculated using experimental calibration method (ECM). Results showed that the average fracture toughness was 485 N/m and 962 N/m, respectively. Finally, through scanning electron micrographs, it was observed that the ply/ply debonding and fibre/matrix debonding were the major fracture mechanisms in DCB specimen. As for ENF specimen, shear fracture dominated the energy dissipation process.

1 Introduction

In recent decades, natural fibres are getting more attention as reinforcements for polymer composites due to their advantages of being low weight, low cost and environmentally friendly (renewable, recyclable and biodegradable) [1]. However, similar to other types of reinforcements, delamination is frequently known as an early failure in composite laminates. Delamination refers to the separation of two adjacent layers of the composite laminates. This would greatly reduce the load carrying capability of the entire composites. Therefore, when using natural fibres as reinforcement, it is of paramount importance to characterise the delamination behaviour in the composites.

In the studies of delamination behaviour of composite laminates, extensive works have been done on pure mode I, II and mixed-mode I+II loadings. Some recent publications on synthetic fibre composites include the research reported in [2–8]. In general, it was observed that the interlaminar fracture toughness increased with the mode II ratio. The highest fracture toughness was obtained under pure mode II loading. This implied that shear mode provided more resistance in delamination and hence more energy was dissipated to initiate the crack propagation. Not only, that the fracture toughness and the variation of the fracture toughness with respect to the mode ratio was dependent on the material, fabrication method, fibre orientation and exposed environment [4–8]. Interlaminar fracture characterisation of polymer reinforced by flax fabrics has been carried out by some researchers [9–12]. For example, the mode I and mode II delamination among different final fibre preform architectures were compared [9]. It was observed that plain weave fabric exhibited the lowest mode I initiation value, G_{IC} (457 N/m) but the highest mode II fracture toughness, G_{IIC} (1872 N/m) compared to other types of architectures. It has been demonstrated that G_{IC} could be improved by incorporating carbon nanotube buckypaper interleaf [10] and through-the-thickness stitching of the flax yarn [11]. Stitching has also shown to have improve G_{IIC} for flax/basalt hybrid composite [12]. In addition, both unstitched and stitched flax/basalt hybrid composites have shown to have higher G_{IIC} as compared to the flax fibre composite itself [12]. Furthermore, upon immersion in the distilled water, G_{IIC} was generally found to have augmented. It was postulated to be due to the increment in the matrix ductility [12]. Based on the literature, it was observed that the delamination characterisation of flax composites was still rather limited. Hence, there is a need for further works to develop better understanding on the delamination behaviour of flax fibre composite.

The objective of this study is to characterise the mode I and mode II interlaminar fracture behaviour of flax fabrics reinforced epoxy composite. The specimens were tested using double cantilever beam (DCB) and three point end notched flexure (ENF) test at initial crack lengths, a_o of 50 mm and 30 mm, respectively. For each test, additional tests at four different initial crack lengths were performed to obtain the compliance plots. The fracture toughness were calculated using experimental calibration method (ECM). Finally, scanning electron micrographs (SEM) of the delaminated surfaces were captured to analyse the failure mechanisms during crack propagation.

2 Materials and Methods

https://doi.org/10.1051/matecconf/201820201002

2.1 Materials

The materials used in this study were 200 g/m^2 twill weave 2/2 flax fabric and 1006 epoxy resin. Firstly, the flax fabrics were cut into 420×420 mm^2 size for 10 plies. Secondly, the epoxy resin was mixed with the hardener at the ratio of 10:6 and stirred gently. Next, the mixture was rolled on the flax fabrics layer by layer. At the mid-plane of the composite, a Teflon film of 15 μm thickness was placed to initiate the pre-crack. The laminate was then cured using vacuum bagging method. Upon completion of the curing process, the composite plate with average thickness of 4.8 mm were cut into specimens at 20 mm width and 200 mm length.

2.2 Interlaminar fracture tests

Double cantilever beam (DCB) and three point end notched flexure (ENF) tests were carried out using a universal testing machine with a 5 kN load cell. Figure 1 shows the test setup of the DCB and ENF tests. For DCB test, the initial crack length, a_o was 50 mm. As for ENF test, specimens were tested at initial crack length, a_o of 30 mm until crack propagation. Three replicates were performed for each test to obtain the average fracture toughness. Besides, for DCB test, additional four specimens at a_o of 30, 35, 40 and 45 mm each were tested. As for ENF test, the additional specimens were loaded at a_o = 20, 25, 35 and 40 mm. The purpose was to obtain the compliance plots for fracture toughness determination using experimental compliance method (ECM). All tests were conducted at crosshead speed of 2 mm/min at ambient conditions.

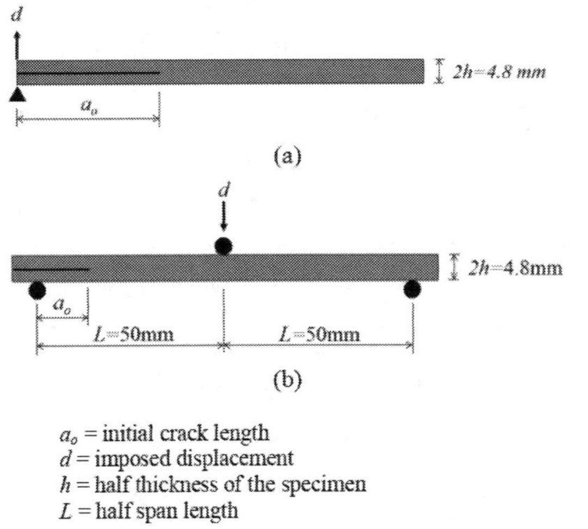

Figure 1. Setup configurations for (a) DCB and (b) ENF tests.

2.3 Fracture surface morphology

To examine the microscopic fracture morphology, the delaminated surfaces of the specimens were first gold coated by 134 Bio-Rad Polaron Division E5100 and images were captured after that using Hitachi S-3400N variable pressure scanning electron microscope (SEM).

3 Results and discussion

3.1 Force-displacement curves

Figure 2 illustrates the experimental force-displacement curves of the flax/epoxy composite under mode I and mode II loadings. Good repeatability was observed for both tests. In addition, it could be seen that all curves increased in a high non-linear manner before the peak load (critical load) was attained. Beyond that, crack was initiated and slight load drop was observed. After that, crack was propagated and the force-displacement curves remained almost plateau.

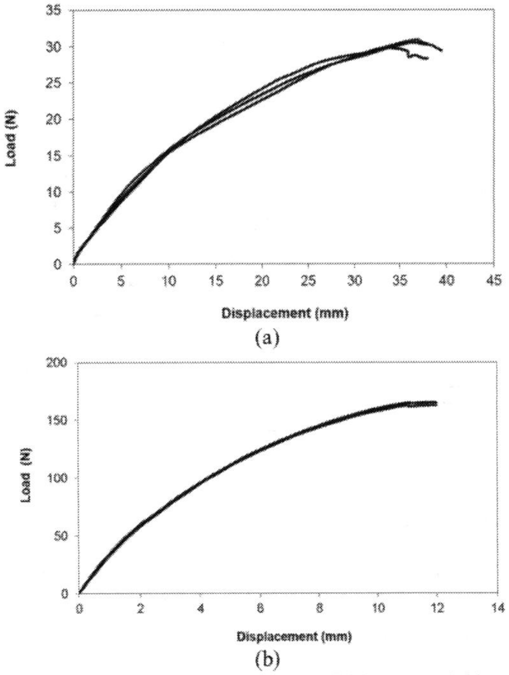

Figure 2. Force-displacement curves of (a) DCB and (b) ENF tests.

3.2 Calculation of interlaminar fracture toughness

The interlaminar fracture toughness, G_C was calculated using experimental calibration method (ECM), which is expressed by:

$$G_C = \frac{P_C^2}{2B} \cdot 3C_2 a_o^2 \qquad (1)$$

In Equation (1), P_C is the peak load from the force-displacement curve, B is the average width of the specimen, C_2 is the coefficient obtained from the following compliance equation:

$$C = C_2 a_o^3 + C_1 \qquad (2)$$

As could be observed from Equation (2), C_1 and C_2 were the coefficients fitted from C versus a_o^3 plot. Figure 3 shows the compliance plots of the flax/epoxy composite at various a_o. From the plot, it could be seen that for DCB test, $C_1 = 1.73\times10^{-1}$ and $C_2 = 3.05\times10^{-6}$. As for ENF test, $C_1 = 1.58\times10^{-2}$ and $C_2 = 5.66\times10^{-7}$. In addition, both plots showed good fit, with $R^2 = 0.98$.

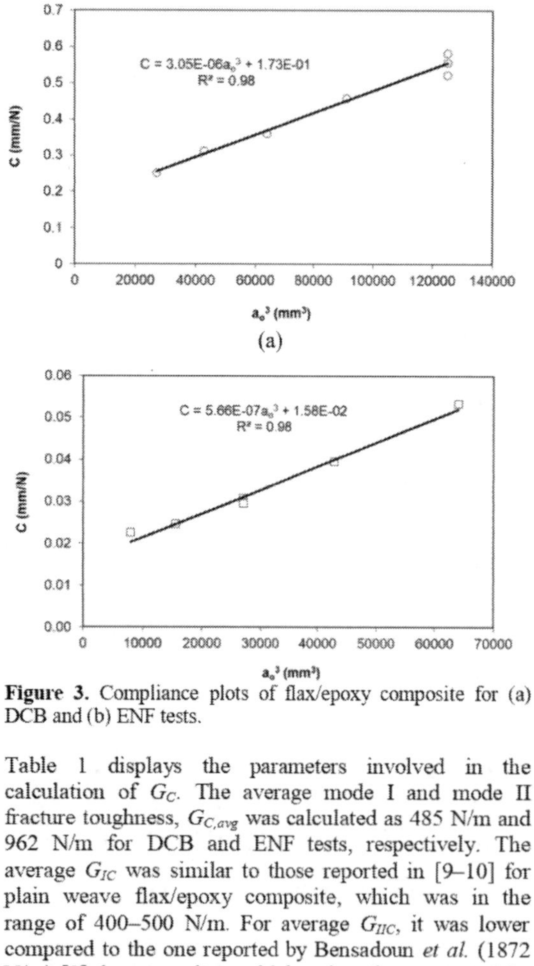

Figure 3. Compliance plots of flax/epoxy composite for (a) DCB and (b) ENF tests.

Table 1 displays the parameters involved in the calculation of G_C. The average mode I and mode II fracture toughness, $G_{C,avg}$ was calculated as 485 N/m and 962 N/m for DCB and ENF tests, respectively. The average G_{IC} was similar to those reported in [9–10] for plain weave flax/epoxy composite, which was in the range of 400–500 N/m. For average G_{IIC}, it was lower compared to the one reported by Bensadoun *et al.* (1872 N/m) [9], however, it was higher than the one tested by Almansour *et al.* (266 N/m) [12]. The possibilities could be due to the different source of flax fabrics, epoxy resin and fabrication method. In addition, the coefficient of variation (C.V) was less than 5 %, which indicated excellent repetition of the results.

Table 1. Calculation of interlaminar fracture toughness of flax/epoxy composite laminates.

B (mm)	P_C (N)	G_C (N/m)	$G_{C,avg}$ (N/m)	C.V (%)
Mode I DCB				
21.93	29.8	464	485	4
21.76	30.5	489		
21.74	30.9	502		
Mode II ENF				
21.25	164.9	977	962	2
21.45	162.6	941		
21.61	165.5	968		

3.3 Morphology study

Figure 4 shows the scanning electron micrographs of the delaminated surfaces of the DCB and ENF specimens. Figure 4(a) shows that significant ply/ply debonding was observed. In addition, fibre/matrix debonding was also noticed. Not only that, there were matrix cracking and small amount of fibrillation on the surface. Some matrix patches were also seen to be attached on the delaminated surface. This could be the reason of having a slight higher G_{IC} compared to the one reported by Chen *et al.* [10]. In their studies, it was reported that the scanning electron micrograph showed clean debonded surface with significant fibre pull-out and bridging [10]. Besides, the mode II fracture surface of flax/epoxy composite (Figure 4(b)) showed shear fracture due to the sliding between the plies was the major fracture mechanism that contributed to energy dissipation process. Furthermore, deformed fibres were noticed, which was an indication of poor interfacial adhesion. According to Almansour *et al.* [12], shear kink bands in flax fibres caused stress concentration in the matrix and fibres. In addition, fibre/matrix debonding and slight amount of fibrillation were observed.

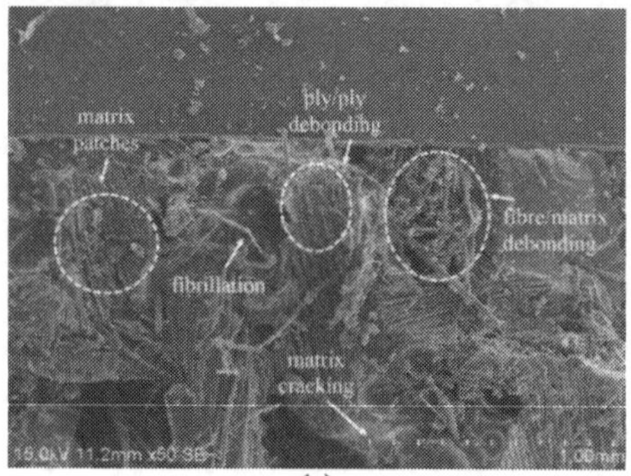

(a)

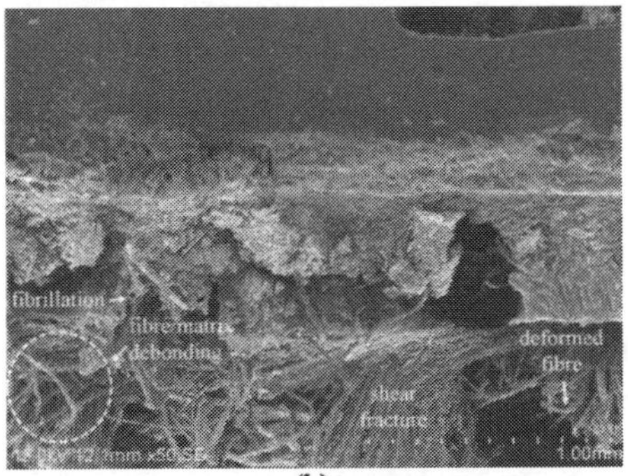

(b)

Figure 4. Scanning electron micrograph of the delaminated surface of flax/epoxy composite under (a) mode I and (b) mode II loadings.

4 Concluding remarks

This study examined the mode I and mode II delamination behaviour of flax fabrics reinforced epoxy composite. Mode I and mode II interlaminar fracture tests were conducted using double cantilever beam (DCB) and three point end notched flexure (ENF) specimens, respectively. The interlaminar fracture toughness were calculated using experimental calibration method (ECM). Significant non-linear behaviour was observed from the force-displacement graphs of both tests. In addition, results showed that the average mode I and mode II fracture toughness were 485 N/m and 962 N/m, respectively. Excellent repetition was found, with maximum coefficient of variation of 4%. Finally, scanning electron micrographs revealed that ply/ply debonding and fibre/matrix debonding were the major fracture mechanisms in the DCB specimen. As for the ENF specimen, shear fracture dominated the fracture process. Deformed fibres were also noticed. This suggested poor fibre/matrix bonding between flax fibre and epoxy resin. Future works could be carried out by improving the interfacial adhesion through fibre surface treatment.

Acknowledgements

This work is supported by Universiti Teknologi Malaysia through Research University Grant (RUG) No. 19H01.

References

1. P.J. Herrera-Franco, A. Valadez-González, Compos. Part A-Appl. S. **35**(3), 339 (2004).
2. M. Johar, H.A. Israr, K.O. Low, K.J. Wong, J. Compos. Mater. **51**(28), 3955 (2017).
3. M. Johar, H.A. Israr, K.J. Wong, M.N. Tamin, *Proceedings of 1st Multidisciplinary Conference on Mechanical Engineering*, 177 (2016).
4. M. Johar, K.J. Wong, M.N. Tamin, *Failure Analysis and Prevention* (InTech, Rijeka, 2017).
5. L.R. LeBlanc, G. LaPlante. Compos. Part A-Appl. S. **81**, 202 (2016).
6. Y. Liu, C. Zhang, Y. Xiang. Compos. Part B-Eng. **82**, 212 (2015).
7. K.J. Wong, X.J. Gong, S. Aivazzadeh, M.N. Tamin, *Structural Composites: Mechanical Properties and Behaviour* (UTM Press, Johor Bahru, 2017).
8. Y. Zhao, W. Liu, L.K. Seah, G.B. Chai, Compos. Part B-Eng. **106**, 332 (2016).
9. F. Bensadoun, I. Verpoest, A.W. Van Vuure, J. Reinf. Plast. Compos. **36**(2), 121 (2017).
10. C. Chen, Y. Li, T. Yu, J. Reinf. Plast. Compos. **33**(20), 1859 (2014).
11. M. Ravandi, W.S. Teo, L.Q.N. Tran, M.S. Yong, T.E. Tay, Mater. Des. **109**, 659 (2016).
12. F.A. Almansour, H.N. Dhakal, Z.Y. Zhang, Compos. Sci. Technol. **154**, 117 (2018).

Effect on the Formation of Magnetite Reduced Graphene Oxide with Controlled Stirring Duration

Xin Hui Yau[1,2], Cheng Seong Khe[1,2], Mohamed Shuaib Mohamed Saheed[1,2] and Chin Wei Lai[3]

[1]*Department of Fundamental and Applied Sciences, Universiti Teknologi PETRONAS, Seri Iskandar, 32610 Perak, Malaysia.*
[2]*Centre of Innovative Nanostructures and Nanodevices (COINN),Universiti Teknologi PETRONAS, 32610 Bandar Seri Iskandar, Perak Darul Ridzuan, Malaysia.*
[3]*Nanotechnology & Catalysis Research Centre (NANOCAT), Institute of Postgraduate Studies (IPS), University of Malaya, 3rd Floor, Block A, 50603 Kuala Lumpur, Malaysia.*

Abstract. Graphene is a promising material due to its fascinating properties, such as mechanical, electronic and thermal properties. Graphene based hybrids materials also have been widely studied due to its wide applications, such as sensors, energy storage and conversion, electronic device and others. The current study presents the synthesis of magnetite-reduced graphene oxide (M-rGO) nanocomposites through in situ chemical synthesis at different stirring durations. This synthesis process involves the redox reaction between the iron(II) salts and graphene oxide (GO) sheets. Various techniques were employed to characterize the synthesized M-rGO nanocomposites. From X-ray diffraction (XRD) results, the crystal structure of M-rGO was found to be independent on the stirring duration. Three magnetite vibrations, D band and G band were observed in Raman spectrum of M-rGO with 24 hours stirring duration. From Fourier transform infrared (FTIR) analysis, M-rGO with 24 hours stirring duration showed the strong intensity of Fe-O vibration. Thus, this indicated that a large amount of magnetite nanoparticles were covered on the surface of rGO sheets. This result is further supported by the morphology of nanocomposites from scanning electron microscopy (SEM) and the elemental analysis (EDX). A monolayer of rGO sheet (C= 33.79 atomic %) with full coverage of magnetite nanoparticles (Fe= 30.20 atomic %) was found for the M-rGO with 24 hours stirring duration. Overall, M-rGO require 24 hours of continuous stirring to ensure full coverage of magnetite nanoparticles on the surface of rGO sheets.

1 Introduction

In 2004, Novoselov and Geim were firstly isolated graphene from graphite through a simple mechanical exfoliation method [1]. Their works had opened a series of opportunities in the realization the usage of graphene in various science and technology due to the remarkable physiochemical properties [2]. Graphene is a one atom thick 2D structure with excellent mechanical, electronic and thermal properties [3, 4]. Graphene oxide (GO) is a precursor material for graphene. GO is recognized as the oxidation product of graphite and it contains carboxyl, carbonyl, epoxyl and hydroxyl groups attached on the graphene sheet and thus can readily disperse in water [5]. Chemical reduction of GO, rGO, involves the removing of attached oxygen rich functional groups from graphene sheet and is an economical route for the production of graphene in large quantities [6]. Owing to the special surface properties, large surface area and layered structured, GO represents an excellent platform for the developing of graphene-based hybrids materials [7]. Recently, these graphene based hybrids materials have been intensively studied due to their wide applications [8, 9].

Iron oxide nanoparticles have also been widely used for incorporation into graphene based materials [10-12]. Among them, magnetite (Fe_3O_4) nanoparticles are the most selective materials for the combination with graphene/graphene oxide based materials [11, 13, 14]. Magnetite is also called iron (II,III) oxide as the formula of magnetite, Fe_3O_4, can be shown as ($FeO \cdot Fe_2O_3$). Magnetite differ from other iron oxides as it contains both Fe^{2+} and Fe^{3+} ions, and these enables magnetite to possess various fascinating properties including superparamagnetic, high coercivity, low curie temperature and high magnetic susceptibility [15, 16]. Therefore, the combination of magnetite nanoparticles with graphene would provide a new and functional hybrid with synergistic interaction between the components.

There are several methods used for the preparation of magnetite graphene based nanocomposites [17-19]. Basically, these methods are classified into two groups, one of which is decorating of Fe_3O_4 on rGO sheet and another one is through chemical reaction of GO with ferrous chloride or ferrous sulphate to form magnetite reduced graphene oxide (M-rGO) nanocomposites. The easier method is the in situ chemical synthesis of M-rGO nanocomposites [20]. The formation of M-rGO nanocomposites involves redox reaction in which the oxidation state of iron (II) is increased to iron (III) while GO is reduced to rGO.

In the present study, we report one of the most convenient methods employed for the synthesis of M-rGO nanocomposites by using in situ chemical synthesis. This study aims to investigate the effect of stirring duration on formation of M-rGO nanocomposites. Hence, the stirring duration from 0.5 hours to 24 hours were examined. The characteristic of the obtained samples were studied by X-ray diffraction (XRD), Fourier transform infrared spectroscopy (FTIR), Raman spectroscopy and field emission scanning electron microscope (FESEM) equipped with energy dispersive X-ray (EDX) detector.

2 Materials and methods

2.1 Materials

Graphite flakes was purchased from Sigma Aldrich. Sulfuric acid (H_2SO_4, 98%), phosphoric acid (H_3PO_4, 85%), potassium permanganate ($KMnO_4$, 99.9%), hydrogen peroxide (H_2O_2, 30%), hydrochloric acid (HCl, 37%), ammonium hydroxide (NH_4OH, 25%) and iron(II) sulphate heptahydrate ($FeSO_4.7H_2O$, 99.5%) were purchased from Merck. All chemicals were of analytical grade and used without any purification. Deionized water was used in all experiments.

2.2 Synthesis of magnetite reduced graphene oxide (M-rGO)

Graphene oxide (GO) in this work was prepared from oxidizing the graphite flakes following the method proposed by Marcano and his co-workers [21]. In a typical method, a 9:1 mixture of H_2SO_4-H_3PO_4 (180:20 mL) was added to a mixture of graphite flakes (1.5 g) and $KMnO_4$ (9.0 g). The mixture was heated to 50 °C and left stirred for three days to ensure complete oxidation of graphite. H_2O_2 was then added to stop the oxidation reaction. The GO was obtained and washed three times with HCl and deionized water through centrifugation until pH 4-5 was achieved. The settlements (GO) obtained were then dried in 40 °C oven to form GO solid. Magnetite reduced graphene oxide (M-rGO) was prepared by in situ chemical method. GO (0.1g) was dispersed in the deionized water with the aid of sonication for 20 mins. NH_4OH (10 mL) was then added to the GO solution until the pH 11-12 was obtained. Subsequently, $FeSO_4$ (0.036 M) was added slowly to the GO solution under the magnetic stirring and stirred at different stirring duration (0.5, 3, 12, 24 h) at room temperature. Black solution was obtained once stirring duration was completed. The solution was then centrifuged and washed three times with deionized water at 4000 rpm for 10 mins. The M-rGO obtained was then dried in 30 °C oven for overnight. The parameter and designations of the prepared samples were shown in Table 1.

Table 1. Synthesis parameter and designations of M-rGO.

Sample	Stirring Duration (hours, h)
M-rGO1	0.5
M-rGO2	3
M-rGO3	12
M-rGO4	24

2.3 Instrumentations

The M-rGO nanocomposites obtained were stored for characterization. Bruker D8 advance diffractometer (XRD) with CuKα radiation was used for identify the crystalline structure of M-rGO samples. The M-rGO pellets were prepared using KBr and the samples were scanned in the range of 500-4000 cm^{-1} to obtain the FTIR spectra by using a Perkin Elmer Spectrum BX Fourier transform infrared spectroscopy (FTIR). Raman spectrum is performed using Horiba Jobin-Yvon HR800 Raman Spectroscopy equipped with a 514.5 nm laser source to examine the change in chemical and crystalline structure during the synthesis process. Surface morphology and the elemental analysis of M-rGO nanocomposites were studied using ZEISS Leo Supra 55 variable pressure field emission scanning electron microscope (FESEM) equipped with energy dispersive X-ray (EDX) detector.

3 Results and discussion

In this experiment, the influence of stirring duration on crystalline nature of the samples was investigated through X-ray diffraction (XRD) analysis. Figure 1(a-d) shows the XRD patterns of M-rGO nanocomposites synthesized at different stirring durations (0.5, 3, 12 and 24 h). In general, M-rGO in Figure 1(a-d) consist of six diffraction peaks which located at $2\theta = 30.0°$, 35.4°, 43.0°, 53.3°, 57.0° and 62.5°, corresponding to the (220), (311), (400), (422), (511) and (440) plane of reflections of magnetite (Fe_3O_4) (JCPDS no. 19-0629). According to Low et al., XRD spectrum of GO has a diffraction peak at $2\theta = 8.59°$ [22]. The disappearance of this peak in the XRD spectra of M-rGO indicated that the oxygen functional groups were removed [22] and rGO was formed. The disappearance of GO reflection peak and the appearance magnetite reflection peaks in the XRD spectra have confirmed that the M-rGO nanocomposites were successfully synthesized through in situ chemical synthesis. The XRD results clearly show that the crystal

 https://doi.org/10.1051/matecconf/201820201003

structure of M-rGO is independent on the stirring duration.

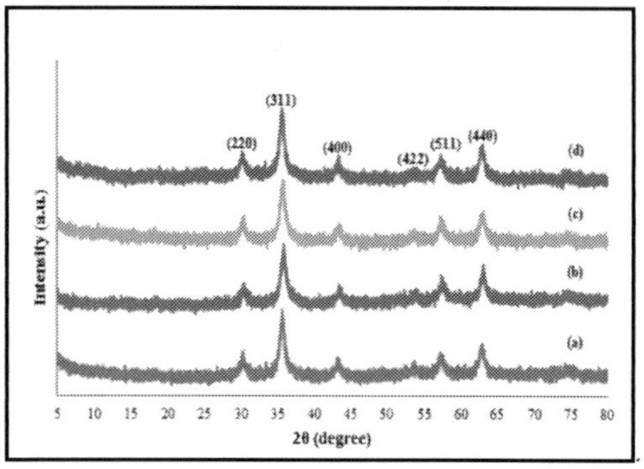

Figure 1. XRD patterns of (a) M-rGO1, (b) M-rGO2, (c) M-rGO3 and (d) M-rGO4.

Fourier transform infrared spectroscopy (FTIR) analysis was carried out to investigate the changes of functional groups of M-rGO nanocomposites that synthesized using different stirring duration. Figure 2(a-d) exhibits the FTIR spectra of various M-rGO nanocomposites. Generally, the FTIR spectra of M-rGO in Figure 2(a-d) have the same characteristic peaks, such as the hydroxyl broad peak (-OH) at about 3420 cm^{-1}, C=C stretching peak at 1623 cm^{-1}, epoxy stretching peak (C-O-C) at 1050 cm^{-1}, and Fe-O lattice vibration at 570 cm^{-1} [23, 24]. However, it can be noticed that the high intensity peak for Fe-O was present for the sample M-rGO4 (Figure 2d). In other words, the M-rGO synthesized at 24 hours continuously stirring has the strongest magnetite Fe-O lattice vibration.

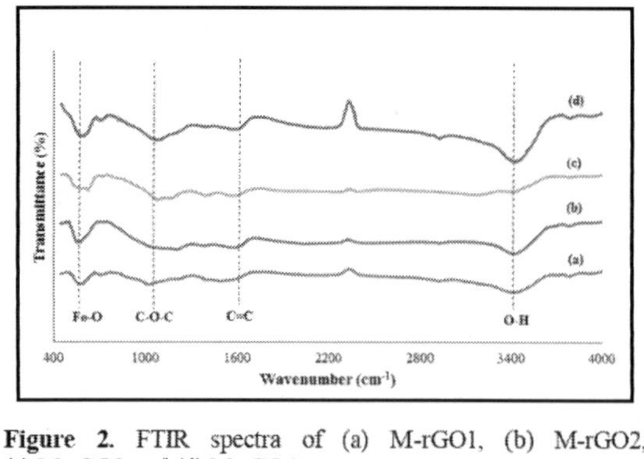

Figure 2. FTIR spectra of (a) M-rGO1, (b) M-rGO2, (c) M-rGO3 and (d) M-rGO4.

Raman spectroscopy is an important tool for the characterization of graphene based nanocomposites. Figure 3 compare the Raman spectra of M-rGO nanocomposites synthesized using various stirring durations. In the M-rGO materials, there are total of five vibrations, including two fundamental vibrations, namely the D and G vibration band, which located at 1350 and 1580 cm^{-1}, respectively, and other vibration bands at 212, 271 and 385 cm^{-1} are originated from magnetite (M). D band is the disorder mode due to the oxidation of graphite, whereas the G band is the first order E_2g mode from sp^2 carbon domain [25]. The Raman spectrum of M-rGO1 as shown in Figure 3(a) has only a D band and a G band. This indicated that a small amount of magnetite nanoparticles are attached on the surface of rGO sheet and thus unable to detect the magnetite vibration in Raman analysis. By contrast, when the M-rGO stirring duration increased from 0.5 h to 24h, the Raman spectrum of M-rGO2 (Figure 3b), M-rGO3 (Figure 3c) and M-rGO4 (Figure 3d), all displayed five vibration peaks in Raman spectra.

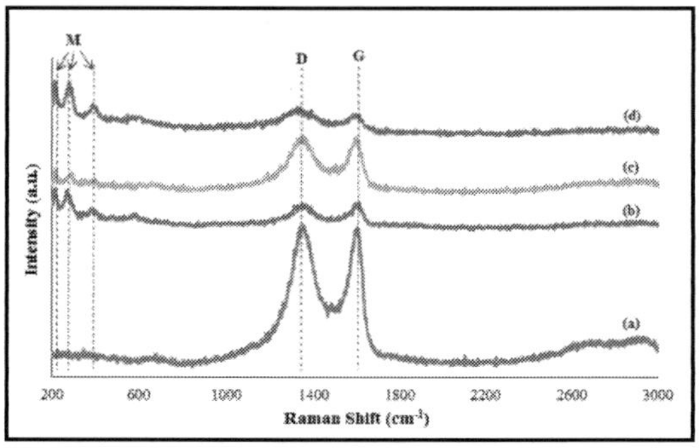

Figure 3. Raman spectra of (a) M-rGO1, (b) M-rGO2, (c) M-rGO3 and (d) M-rGO4.

Field emission scanning electron microscopy (FESEM) was carried out in order to investigate the morphology of the synthesized M-rGO samples. Figure 4(a-d) shows the FESEM images of M-rGO samples after various stirring durations (0.5, 3, 12 and 24h). FESEM images in Figure 4(a) shows that a layered of rGO sheet with little amount of magnetite nanoparticles attached on it. This result again proved that 0.5 hours stirring duration is not enough for the formation of M-rGO nanocomposites. When the stirring duration increased to 3 hours (Figure 4b) and 12 hours (Figure 4c), more magnetite nanoparticles were attached on rGO sheet. Meanwhile, M-rGO with 24 hours continuously stirring as shown in figure 4(d) has a fully coverage of magnetite nanoparticles on the monolayer of rGO sheet. The FESEM results clarify that as the stirring duration increase, more magnetite nanoparticles are attached on the rGO sheet.

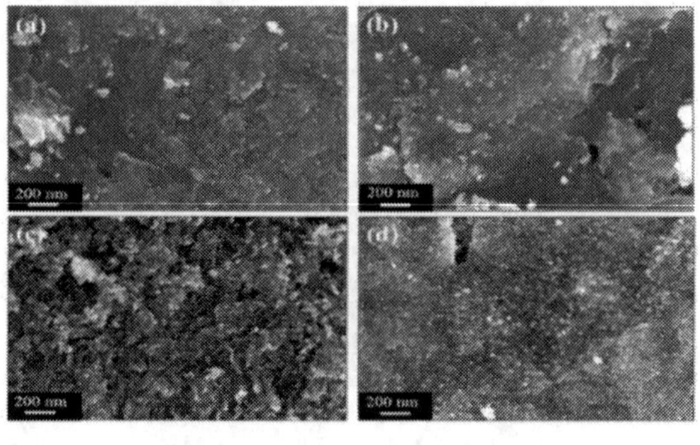

Figure 4. FESEM images of (a) M-rGO1, (b) M-rGO2, (c) M-rGO3 and (d) M-rGO4.

Table 2 indicated the composition of synthesized M-rGO samples obtained from the energy dispersive X-ray (EDX) detector from FESEM instrument. From EDX results, the M-rGO4 has the lowest amount of carbon, C with 33.79 atomic % and highest amount of iron, Fe with 30.20 atomic %. The ratio of iron to carbon (Fe/C) was calculated and M-rGO4 achieved the highest Fe/C ratio (Fe/C= 0.89). Undeniably, M-rGO with 24 hours continuously stirring has the higher amount of magnetite nanoparticles attached on the rGO sheet and the lowest amount of carbon detector suggested that the magnetite nanoparticles were well covered on the surface of rGO sheet.

Table 2. Summarize of EDX analysis (in atomic %) of M-rGO.

Sample	C (%)	O (%)	Fe (%)	Fe/C ratio
M-rGO1	37.02	46.81	16.17	0.44
M-rGO2	38.35	45.19	16.46	0.43
M-rGO3	45.82	36.29	17.89	0.39
M-rGO4	33.79	36.01	30.20	0.89

Based on the results obtained, a brief and plausible interaction mechanism between the iron(II) salt and GO in aqueous was proposed to explain the formation of M-rGO nanocomposites. Both iron(II) salt and GO in aqueous solution interact through coordinate (dative covalent) bonding. Iron(II) salt in aqueous solution tend to act as cations, which are Fe^{2+}. Cations are electron deficient with unoccupied orbitals. Meanwhile, the GO surface consist a lot of electron rich oxygen atoms, such as hydroxyl (-OH) and epoxy (-O-) groups with lone pairs of electrons. The coordinate bond is formed between the Fe^{2+} and oxygen atoms of GO, thus forming the M-rGO nanocomposites.

4 Conclusion

In conclusion, the magnetite reduced graphene oxide (M-rGO) nanocomposites were successfully synthesized through in situ chemical synthesis. The longer the stirring duration, more magnetite nanoparticles were attached on the surface of rGO sheet. M-rGO4 or M-rGO with 24 hours of continuously stirring is the best sample among others as it contain low amount of carbon atom (33.79 atomic%) and high amount of iron atom (30.20 atomic%). A monolayer of rGO sheet with the fully coverage of magnetite nanoparticles was obtained for the M-rGO samples with 24 hours stirring duration.

4 Acknowledgments

Authors acknowledge Yayasan Universiti Teknologi PETRONAS (YUTP) Grant (YUTP 0153AA-E40) and Graduate Assistantship program from the Universiti Teknologi PETRONAS for providing financial assistance.

References

1. K. S. Novoselov, A. K. Geim, S. V. Morozov, D. Jiang, Y. Zhang, S. V. Dubonos, I. V. Grigorieva and A. A. Firsov, Science **306** (5696), 666-669 (2004).
2. K. S. Novoselov, V. Fal, L. Colombo, P. Gellert, M. Schwab and K. Kim, Nature **490** (7419), 192 (2012).
3. A. K. Geim and K. S. Novoselov, Nat. Mater. **6** (3), 183 (2007).
4. X. Huang, Z. Yin, S. Wu, X. Qi, Q. He, Q. Zhang, Q. Yan, F. Boey and H. Zhang, Small **7** (14), 1876-1902 (2011).
5. D. Konios, M. M. Stylianakis, E. Stratakis and E. Kymakis, J. Colloid Interface Sci. **430**, 108-112 (2014).
6. D. Li, M. B. Müller, S. Gilje, R. B. Kaner and G. G. Wallace, Nat. Nanotechnol. **3** (2), 101 (2008).
7. M. Şinoforoğlu, B. Gür, M. Arık, Y. Onganer and K. Meral, RSC Adv. **3** (29), 11832-11838 (2013).
8. S. Stankovich, D. A. Dikin, G. H. Dommett, K. M. Kohlhaas, E. J. Zimney, E. A. Stach, R. D. Piner, S. T. Nguyen and R. S. Ruoff, Nature **442** (7100), 282 (2006).
9. M. Pumera and C. H. A. Wong, Chem. Soc. Rev. **42** (14), 5987-5995 (2013).
10. B. Saiphaneendra, T. Saxena, S. A. Singh, G. Madras and C. Srivastava, J. Environ. Chem. Eng. **5** (1), 26-37 (2017).
11. R. Hatel, M. Goumri, B. Ratier and M. Baitoul, Mater. Chem. Phys. **193**, 156-163 (2017).
12. J. Liu, H. Cao, J. Xiong and Z. Cheng, Cryst. Eng. Comm. **14** (16), 5140-5144 (2012).
13. M. Liu, C. Chen, J. Hu, X. Wu and X. Wang, The J. Phys. Chem. C **115** (51), 25234-25240 (2011).
14. V. Chandra, J. Park, Y. Chun, J. W. Lee, I.-C. Hwang and K. S. Kim, ACS Nano **4** (7), 3979-3986 (2010).
15. W. Wu, Q. He and C. Jiang, Nanoscale Res. Lett. **3** (11), 397 (2008).
16. L. Blaney, Lehigh Rev. **15**, 33-81(2007).
17. V. Hoan, N. Thi, A. Thu, N. Thi, H. V. Duc, N. D. Cuong, D. Quang Khieu and V. Vo, J. Chem. **2016** (2016).
18. Y. Sun, W. Zhang, H. Yu, C. Hou, D.-s. Li, Y. Zhang and Y. Liu, J. Alloys Compd. **638**, 182-187 (2015).
19. A. Meidanchi and O. Akhavan, Carbon **69**, 230-238 (2014).
20. P. S. Teo, H. N. Lim, N. M. Huang, C. H. Chia and I. Harrison, Ceram. Int. **38** (8), 6411-6416 (2012).
21. D. C. Marcano, D. V. Kosynkin, J. M. Berlin, A. Sinitskii, Z. Sun, A. Slesarev, L. B. Alemany, W. Lu and J. M. Tour, ACS Nano **4** (8), 4806-4814 (2010).
22. F. W. Low, C. W. Lai and S. B. A. Hamid, Ceram Int. **41** (4), 5798-5806 (2015).
23. J. Liang, H. Li, J. Yan and W. Hou, Energy & Fuels **28** (9), 6172-6178 (2014).

https://doi.org/10.1051/matecconf/201820201003

24. H. Wang, J. Liu, H. Xu, Z. Ma, W. Jia and S. Ren, RSC Advances **6** (108), 106297-106307 (2016).
25. K. N. Kudin, B. Ozbas, H. C. Schniepp, R. K. Prud'Homme, I. A. Aksay and R. Car, Nano Lett. **8** (1), 36-41 (2008).

Bond Layer Properties and Geometry Effect on Interfacial Thermo-mechanical Stresses in Bi-material Electronic Packaging Assembly

D.Sujan, L. Vincent and Y. W.Pok

Faculty of Engineering and Science, Curtin University Malaysia
CDT 250, 98009 Miri, Sarawak, Malaysia

Abstract. Thermo-mechanical mismatch stress is one of the reasons for mechanical as well as functional failure between two or more connected devices. In electronic packaging, two or more plates or layers are bonded together by an extremely thin layer. This thin bonding layer works as an interfacial stress compliance which is expected to alleviate the interfacial stresses between the layers. Therefore, it is very important to identify the suitable interfacial bonding characteristics for reducing the interfacial thermal mismatch stresses to maintain the structural integrity. This research work examines the influences of bond layer properties and geometry on the interfacial shearing and peeling stresses in a bi-material assembly. In this study a closed form model of bi-layered assembly is used with the up-to-date bond layer shear stress compliance expression. The key bond layer properties namely Young's modulus, coefficient of thermal expansion, Poisson's ratio, and physical parameters like temperature and thickness are considered for interfacial stress evaluation. It is observed that the Young's modulus, the thickness and the temperature of the bond layer have significant influence on the interfacial shearing and peeling stress. The results obtained are likely to be useful in designing bond layer properties in microelectronics and photonics applications.

1 Introduction

Interfacial thermal stresses are induced due to the mismatches of thermal, stiffness and other properties caused by the dissimilar materials. This mismatch of properties is often unavoidable due to the functional requirement of the materials in the package. Unfortunately this induced thermo-mechanical mismatch stress is one of the main causes of mechanical and functional failure in electronic and photonic devices. As the power requirement of the devices increases, the problem with thermal mismatch assumes a major cause of failure in electronic packages [1-8].

Nevertheless, in electronic packaging, two or more materials are bonded together by an extremely thin interfacial stress compliant attachment. This bond material can provide appreciable buffering effect which can lead to a substantial relief in thermally induced stresses in various adhesively bonded or soldered assemblies in electronics and photonics [8-9] Selection of bond material properties and thickness should also be addressed at the physical design stage for optimum mechanical performance of the device.

Three main approaches can analyze thermal stresses in layered structure namely 2-D elasticity solution, elementary beam theory and finite element (FE) analysis [10]. However, beam theory is very popular among the researchers due to its simplicity. In beam theory, it considers a unit strip is cut parallel to the plane and then analyzes the stresses in that strip. The analysis is even more simplified by replacing the stresses on the cross-section by equivalent resultant forces and moments [1].

Suhir [11-12] studied the magnitude and distribution of stresses in adhesively bonded bi-material assemblies, with consideration of the attachment compliance. Compliance attachments, providing a strain buffer between thermally mismatched materials, can offer substantial stress relief, and for this reason, are considered in many cases as an acceptable solution to the thermal mismatch problem. Suhir showed that for relatively small assemblies, compliant attachment could indeed result in substantial stress relief in both the adherents and the adhesives.

Sujan proposed a model for shearing and peeling stresses at the interface of bi-material assembly which satisfies the interface compatibility by matching of strains at the interface rather than the actual displacements as in Suhir's work [1, 6, 13-15]. As a consequence, an unnecessary integro-differential equation was avoided by solving only a simpler second order differential equation.

Suhir [14] defined interfacial shear stress compliance for bond layer K_0 as $2t_0/G_0$, where t_0 is the thickness and

G_0 is the shear modulus of rigidity of the bond material. Schmidt [5] used the relationship for K_0 as h_0/G_0. However, Schmidt did not show any derivation for K_0. Schmidt also carried out a numerical solution for a bi-material assembly with bond layer where shear stress is compared for different K_0 values i.e. $2t_0/3G_0$, $t_0/3G_0$, and t_0/G_0. Sujan proposed shear stress compliance with complete derivation eliminated the contradiction of using different compliance expressions by earlier researchers [4, 6, 15].

In this study a closed form model of bi-layered assembly is used with the up-to-date bond layer shear stress compliance expression [4]. The key bond layer properties namely Young's modulus, coefficient of thermal expansion, Poisson's ratio, and physical parameters like temperature and thickness are considered for interfacial stress evaluation.

2 Analytical model

The analytical model used in this project is developed using second order differential equation which does not involved any integro-differential equation. Figure 1(a) shows an arbitrary location of the model in 2-dimensional form and Figure 1(b) shows the free body diagram of the full length of the model. The model is considered to be a unit of width in a direction perpendicular to the plane of the paper and the forces and moments are defined with respect to the unit width.

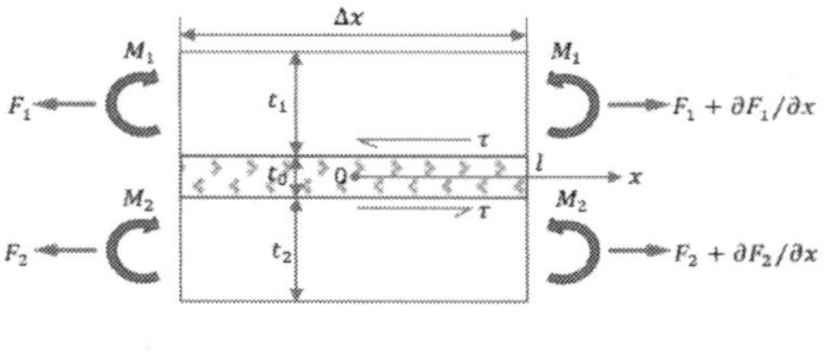

(a)

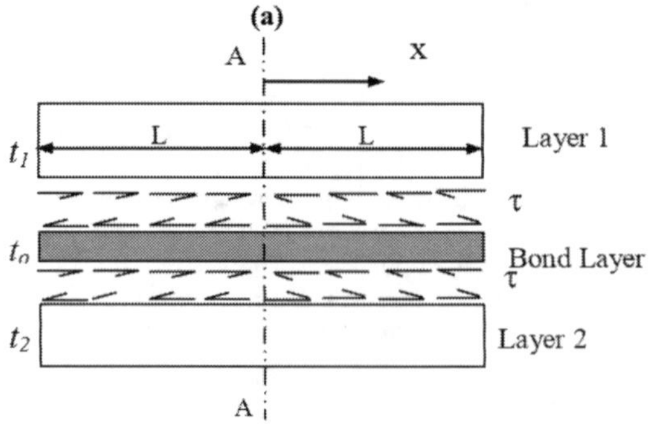

(b)

Figure 1. (a) Bi-material assembly and (b) free-body diagram of the model [4, 6, 15]

Symbols used in this paper:

Material or layer number, i=1, 2; E = Young's modulus (N/m^2); t_i = Thickness (m); ΔT=120°C
α_i = Coefficient of thermal Expansion (1/°C);
ν_i = poison's ratio; R = Radius of curvature; Shear modulus of rigidity, $G_i = \dfrac{E_i}{2(1+\nu_i)}$ (N/m^2); Flexural rigidity, $D_i = \dfrac{E_i t_i^3}{12(1-\nu_i^2)}$ (kg.m); Effective flexural rigidity, $D = D_1 + D_2$ (kg.m); Axial compliance, $\lambda_i = \dfrac{1}{E_i t_i}$ (m/kg); Shear compliance for top and bottom layer, $K_i = \dfrac{t_i}{3G_i}$ (m^3/N); Shear compliance for the bond layer, $K_0 = \dfrac{t_0}{G_0}$ (m^3/N)

The solution is based on the assumptions as follows:

1. Thickness of the layered assembly is relatively small.
2. Each layer can be regarded as Bernoulli beam
3. Spherically bending thin plate is acted in each layer.
4. No external force acting among them.
5. Axial force due to thermal loading varies along the length and full shear length in the interface bonded layers.
6. Adhesive layer (solder bond) is very thin compared to the top and bottom layers

The shear stress τ(x) is given by,

$$\tau = \frac{\Delta T(\alpha_1 - \alpha_2)}{K\kappa\, cosh(\kappa L)} sinh(\kappa x) \qquad (1)$$

The peeling stress P(x) expression is given by,

$$P = \frac{(t_1 D_2 - t_2 D_1)}{2D} \frac{\Delta T(\alpha_1 - \alpha_2)}{K cosh(\kappa L)} cosh(\kappa x) \qquad (2)$$

Where $t = t_1 + t_2 + 2t_0$, $\lambda = \lambda_1 + \lambda_2 + \dfrac{t(t_1 + t_2)}{4D}$,

and $K = K_1 + K_2 + K_0$ and $\kappa = \sqrt{\dfrac{\lambda}{K}}$

3 Numerical example:

An arbitrary bi-layered package system was used in this study. **Table 1** shows the parameters used to calculate the interfacial stresses by using the analytical approach. The length of the assembly is $2L = 0.005$ m. The temperature changes, ΔT is taken at 120°C in this computation.

Table 1. Material properties and dimension

Material	Young's Modulus, *E* (GPa)	CTE, α (°C), 10^{-6}	Poisson's Ratio, ν	Thickness *t*, mm
Die	110	2.9	0.31	0.40
Die attach	141	22	0.28	0.20
Bond layer	10-70	5-25	0.25-0.33	0.04 - 0.08

4 Results and discussion

The shearing stress and peeling stress are calculated using eq. (1) and eq. (2) respectively. The results are plotted in Figure 2 to Figure 5 and Table 3 to Table 4.The interfacial stresses are tabulated from x/L = 0.91-1.0, since the stress values are significantly high in the vicinity of the edge. The maximum shearing stress and peeling stress are recorded at the free end ($x/L = 1.0$) as expected.

4.1 Young's modulus (stiffness) effect

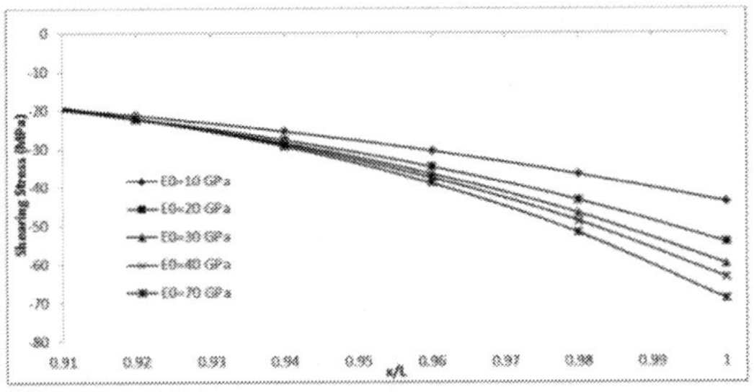

Figure 2. Shear Stress along the interface with Young's Modulus as parameter

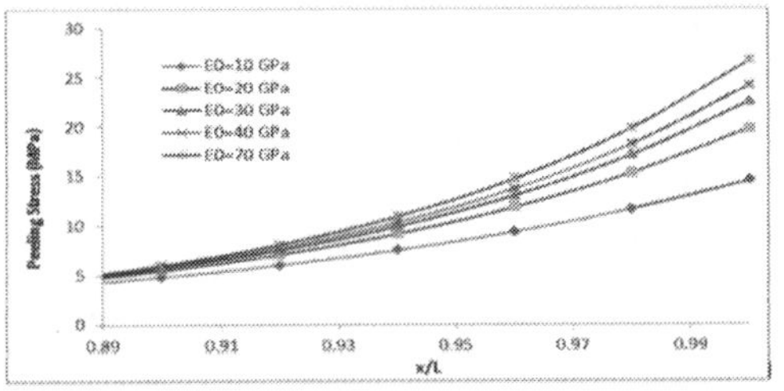

Figure 3. Peeling stresses along the interface with Young's Modulus as parameter

Figure 2 and **Figure 3** represent shearing and peeling stresses for different values of Young modulus along the interface. The Young modulus of bond layer, E_0 is varied from 10 to 70 MPa at the interval of 10, 20, 30, 40, and 70 MPa. From the Figure 2 and 3 above, it can be noticed that the shearing stress is tensile while peeling stress is compressive in nature along the interface as suggested in theory. It is observed that the shearing stress and peeling stress decreased due to the compliant (lower young's modulus) bond layer effect at any identical location at the interface. Particularly, near the vicinity of the free end, the differences of thermo-mechanical stress between highest modulus and lowest modulus of bond layer were significant, which is 76.1% for shearing stress and 84.5% for peeling stress. Thus, it indicates that a more compliant bond will likely to result in smaller interfacial stress compared to a stiffer bond. At this point, it is worth mentioning here that a compliant bond is likely to generate smaller interfacial stress but it is easier to deform. On the other hand, a stiffer bond is likely to results in higher stresses but it is more difficult to deform [3].

4.2 Thickness effect

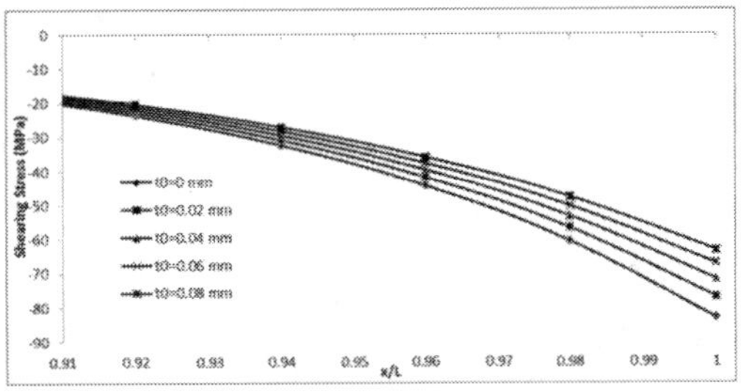

Figure 4. Shear stresses along the interface of different bond layer thickness.

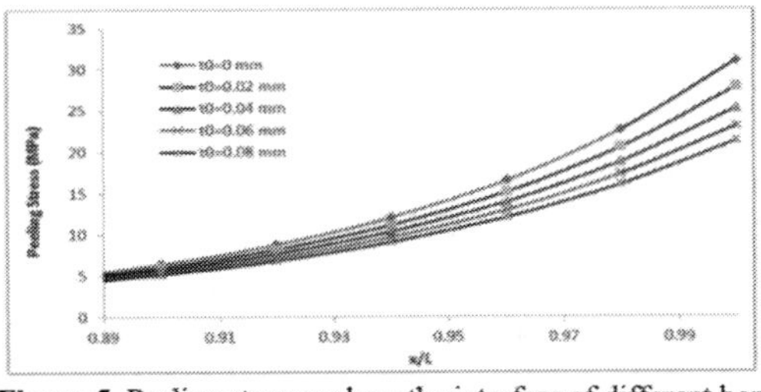

Figure 5. Peeling stresses along the interface of different bond layer thickness.

Figure 4 and Figure 5 compared shearing stress and peeling stress respectively with bond layer thickness, h_0 as a parameter. The results are plotted only from x/L= 0.91 to 1 to represent significant effects of bond layer thickness near the vicinity of the free end.

It is evident from Figure 4 and Figure 5 that shearing and peeling stresses decreased with the increase of compliant bond layer thickness at any identical location along the interface. For instance, at location x/L =1, the shearing stress recorded as 157.9 MPa, 120.8 MPa, 109.5 MPa, 100.4MPa, and 93.06MPa with bond layer thicknesses of h_0=0 mm, 0.02 mm, 0.04 mm, 0.06 mm, and 0.08 mm respectively. Similar trend of variation is also observed in Figure 5 for the case of peeling stresses.

For both interfacial stresses distribution, the shear stress and peeling stress were observed to be reduced along the length when the thickness of bond layer was being increased. The increased bond layer thickness would generated a surrogate layer acted for shear stress absorption along the interface.

Therefore, a higher value of bond layer thickness would be suggested in the flip-chip packaging design. However, the increased bond thickness was seem to be another concern for the increased overall cost of material.

4.3 Poisson's ratio effect

Table 2. Stresses at location x/L =1 for different Poisson's ratio

Poisson's Ratio, v	**0.25**	**0.27**	**0.29**	**0.31**	**0.33**
Shear Stress (MPa)	-43.9	-43.7	-43.5	-43.2	-43.0
Peeling Stress (MPa)	29.9	29.8	29.7	29.6	29.5

Table 2 represents the shear and peeling stress values at location x/L = 1 (edge) with Poisson's Ratio varied in the range of 0.25-0.33. It can be observed that the both shear and peeling stress changes very slightly due to the variation of Poisson's Ratio. Thus, Poisson ratio of bond layer do not play significant role in reducing interfacial stresses in layered structure. From the above observation it can be concluded that the effect of Poisson ratio in bond layer may not be essential to consider in predicting stresses development at the interface.

4.4 Thermal expansion coefficient effect

From eq. (1) and Eq. (2), it can be observed that the coefficient of thermal expansion of bond layer term is absent or has no effect in shearing and peeling stress expressions for with bond layer consideration. This is due to the fact that the bond layer is extremely thin compared to the two adhere layers. Therefore, it does not contribute much in terms of expansion and compression of the overall package. Therefore, it can be concluded that the coefficient of thermal expansion of the bond layer insignificant contribution in predicting stresses development at the interface.

4.5 Temperature Effect

Table 3. Stresses at location x/L =1 for different temperature

Temperature, °C	**60**	**80**	**100**	**120**	**140**
Shear Stress (MPa)	-31.7	-42.3	-52.9	-63.5	-74.1
Peeling Stress (MPa)	17.4	23.2	29.0	34.8	40.6

From eq. (1) and eq. (2), it can be observed that the bond layer term is also absent in shearing and peeling stress expressions for with bond layer consideration. This is due to the fact that the bond layer is extremely thin compared to the two adhere layers. Therefore, it can be logically assumed that the bond layer temperature will be same as the overall package. Nevertheless, Table 3 clearly indicates that the temperature is a very significant factor in stress development in the interface. For instance, shear stress increases 233% for an increment of temperature from 60°C to 120°C at the edge of the package. Therefore, it can be concluded that the temperature should be as low as possible during manufacturing and operation of electronic packages to avoid mechanical and functional failure.

5 Conclusions

This research work examined the influences of bond layer properties and geometry on the interfacial shearing and peeling stresses in a bi-material assembly. In this study a closed form model of bi-layered assembly was used with the up-to-date bond layer shear stress compliance expression. The key bond layer properties namely Young's modulus, coefficient of thermal expansion, Poisson's ratio, and physical parameters like temperature and thickness were considered for interfacial stress evaluation. It is observed that the Young's modulus, the thickness and the temperature of the bond layer have significant influence on the interfacial shearing and peeling stress. The shearing stresses and peeling stresses decreased considerably at the interface with the increase of bond layer Young Modulus and thickness. On the other hand, Bond layer coefficient of thermal expansion and Poisson ratio show almost no significant effect on the interfacial shearing stress and peeling stress along the interface in a bi-material assembly. The results obtained are likely to be useful in designing bond layer properties in microelectronics and photonics applications.

Acknowledgement

The authors thankfully acknowledge the research facilities and infrastructure provided by the Curtin University Malaysia and the financial assistance provided by the Ministry of Higher education (Malaysia) under FRGS scheme, Grant reference Code: FRGS/2/2013/SG06/CURTIN/02/1

References

1. Timoshenko. S. (1925). Analysis of Bi-Metal Thermostats. Journal of the Optical Society of America.11, pp. 233-255
2. Tummala, R. (2001). Fundamentals of microsystems packaging . McGraw Hill Professional.
3. Moore, G. E. (1998). Cramming more components onto integrated circuits. Proceedings of the IEEE, 86(1), 82-85. doi:10.1109/JPROC.1998.658762
4. Sujan, D., Woldemichael, D. E., Murthy, M. V. V., & Seetharamu, K. N. (2011). Effect of bond layer on

bimaterial assembly subjected to uniform temperature change. Journal of Electronic Packaging, Transactions of the ASME, 133(4) doi:10.1115/1.4005294
5. Brown, W. D. (1998). Advanced Electronic Packaging with Emphasis on Multichip Modules. ed.
6. Sujan, D., Pang, X. B., Rahman, M. E., & Reddy, M. M. (2014). Performance of solder bond on thermal mismatch stresses in electronic packaging assembly doi:10.4028/www.scientific.net/MSF.773-774.242
7. Huang, Y.W., Lin, Y. M., Zhan. C.J. (2013). Assembly of 3D Chip Stack with 30μm-Pitch Micro Interconnects Using Novel Arrayed-Particles Anisotropic Conductive Film. Proceedings of the 2013 Electronic Components and Technology Conference, pp. 71-76.
8. Suhir, E. (2018). Analytical thermal stress model for a typical flip-chip (FC) package design. Journal of Materials Science: Materials in Electronics, 29(4), 2676-2688. doi:10.1007/s10854-017-8194-6
9. Suhir, E. (2014). Thermal stress in through-silicon-vias: Theory-of-elasticity approach. Microelectronics Reliability, 54(5), 972-977. doi:10.1016/j.microrel. 2014.01.004
10. Gao, L., Chen, X., & Gao, H. (2015). Interfacial thermal stresses in ACF bonding assembly. Microelectronics Reliability, 55(7), 1089-1096. doi:10.1016/j.microrel.2015.04.004
11. Suhir, E. (1989). Interfacial stresses in bimetal thermostats. Journal of Applied Mechanics, Transactions ASME, 56(3), 595-600. doi:10.1115/1.3176133
12. Suhir, E. (1986). CALCULATED THERMALLY INDUCED STRESSES IN ADHESIVELY BONDED AND SOLDERED ASSEMBLIES. Paper presented at the 383-392.
13. Sujan, D., Murthy, M. V. V., Seetharamu, K. N., & Hassan, A. Y. (2005). Complete model for interfacial stresses of a two layered structure. Paper presented at the Proceedings of the 6th International Conference on Thermal, Mechanical and Multi-Physics Simulation and Experiments in Micro-Electronics and Micro-Systems - EuroSimE 2005, , 2005 454-461. doi:10.1109/ESIME.2005.1502848
14. Sujan, D., Murthy, M. V. V., & Seetharamu, K. N. (2011). Improved tri-layered interfacial stress model with the effect of different temperatures in the layers. Archive of Applied Mechanics, 81(5), 561-568. doi:10.1007/s00419-010-0437-4
15. Sujan, D., Murthy, M. V. V., Seetharamu, K. N., & Hassan, A. Y. (2005). Engineering model for interfacial stresses of a heated bimaterial structure with bond material used in electronic packages. Journal of Microelectronics and Electronic Packaging, 2(2), 132-141. doi:10.4071/1551-4897-2.2.132

Material Selection for Interfacial Bond Layer in Electronic Packaging

D.Sujan , L. Vincent and Y. W. Pok

Faculty of Engineering and Science, Curtin University Malaysia
CDT 250, 98009 Miri, Sarawak, Malaysia

Abstract. In electronic packaging, typically two or more thin dissimilar plates or layers are bonded together by an extremely thin adhesive bond layer. Electronic assemblies are usually operated under high power conditions which predictably produces a high temperature environment in the electronic devices. Therefore, thermal mismatch shear and peeling stress inevitably arise at the interfaces of the bonded dissimilar materials due to differences in Coefficient of Thermal Expansion (CTE) typically during the high temperature change in the bond process. As a result, delamination failure may occur during manufacturing, machining, and field use. As such, these thermo-mechanical stresses play a very significant role in the design and reliability of the electronic packaging assembly. Consequently, critical investigations of interfacial stresses under variable load conditions in composite structure can result in a better design of electronic packaging with higher reliability and minimize or eliminate the risk of functional failure. In order to formulize bond material selection, analytical studies are carried out in order to study the influence of bond layer parameters on interfacial thermal stresses of a given package. These parameters include Coefficient of thermal expansion (CTE), poison's ratio, temperature, thickness, and stiffness (compliant and stiff) of the bond layer. From the study, stiffness and bond layer thickness are identified as the key parameters influencing interfacial shearing and peeling stresses. The other parameters namely CTE, poisons ratio has shown insignificant influence on interfacial stresses due to the very thin section of bond layer compared to the top and bottom layers. The results also show that the interfacial stresses increases proportionally with the increase of temperature in the layers. Therefore, it is very important that the temperature is maintained as low as possible during the chip manufacturing and operating stages. Since only two parameters namely stiffness and bond layer thickness are identified as the key parameters, the interface thermal mismatch stresses can be reduced or eliminated by controlling these two parameters only. Therefore the identification of suitable bond layer parameters selection with reasonable accuracy is possible even without performing optimization process. Finally, this paper proposes a Metal Matrix Composite (MMC) bond material selection approach using rule of mixture material design. The outcome of this research can be seen in the forms of practical and beneficial tools for interfacial stress evaluation and physical design and fabrication of layered assemblies. The Engineers can utilize this research outcome in conjunction with guidelines for electronic packaging under variable thermal properties of layered composites.

1 Introduction

In electronic packaging, typically two or more thin dissimilar plates or layers are bonded together by an extremely thin adhesive bond layer. Electronic assemblies are usually operated under high power conditions which predictably produces a high temperature environment in the electronic devices. Therefore, thermal mismatch shear and peeling stress inevitably arise at the interfaces of the bonded dissimilar materials due to differences in Coefficient of Thermal Expansion (CTE) typically during the high temperature change in the bonding process. As a result, delamination failure may occur during manufacturing, machining, and field use. As such, these thermo-mechanical stresses play a very significant role in the design and reliability of the electronic packaging assembly. Consequently, critical investigations of interfacial stresses under variable load conditions in composite structure can result in a better design of electronic packaging with higher reliability and minimize or eliminate the risk of functional failure [1-10].

The selection of electronic packaging materials (including bond layer) for optimum performance under thermal loading is a challenging task to minimize or eliminate the risk of structural failure. Therefore, there is a compulsion to understand the mechanism of interfacial stresses development in electronic packaging [7, 11]. Bi-layered analytical model for interfacial shear and peeling stress has been developed by Suhir [12] using integro-differential equation. Suhir expressed the compatibility condition at the interface in terms of strain. Sujan et al [15, 7, 13-15] had developed a simple solution using a second order differential equation approach. Sujan used the compatibility at the interface in terms of displacement. For both Suhir and Sujan, Models are mathematically equivalent and thus lead to the same solution. The effect of bond layer properties and geometry would be another important concern in developing an optimized solution in electronic packaging.

The conventional leaded bond material is found to have poor thermal mismatch resistance. Therefore, in recent years, there are novel efforts in the new development on lead free bond materials. On the other hand, particulate reinforced light Metal Matrix Composites (MMC) have shown great promise due to their outstanding tailor-made mechanical and physical properties.

This research is proposed to formulate a lead free bond material selection approach in electronic packaging to estimate the best possible mechanical and physical properties of the bond material in relation to the packaging materials in order to minimize thermal mismatch stresses. Subsequently, MMC bond material selection strategy is developed by using the rule of mixture (ROM) design approach [16]. ROM is a method which is based on the assumption that a composite property is the volume weighted average of the total constituents of the material. MMC approach can be utilized later (which is out of the scope of this project) in order to synthesize the tailored sustainable bond material recommended by the reference model.

The outcome of this research is in the forms of practical and beneficial tools for interfacial stress evaluation and physical design of layered assemblies. The Engineers can utilize this reference model in conjunction with guidelines for electronic packaging under variable thermal properties of layered composites.

In this study a closed form model of bi-layered assembly is used with the up-to-date bond layer shear stress compliance expression [5]. The key bond layer properties namely Young's modulus, coefficient of thermal expansion, Poisson's ratio, and physical parameters like temperature and thickness are considered for interfacial stress evaluation.

2 Analytical Model

The same analytical model which has been used in paper 1 (Title: Bond layer properties and geometry effect on interfacial thermo-mechanical stresses in bi-material electronic packaging assembly) in this conference is utilized for bond material selection, and design approach. Therefore, only brief discussion of the model and the findings of the paper 1 will be discussed here for the process. Since maximum stressed are recorded at the location of x/L =1 (edge), only results related to this location will be considered.

Figure 1(a) shows an arbitrary location of the model in 2-dimensional form. Figure 1(b) shows the free body diagram of the full length of the model. The model is considered to be a unit of width in a direction perpendicular to the plane of the paper and the forces and moments are defined with respect to the unit width.

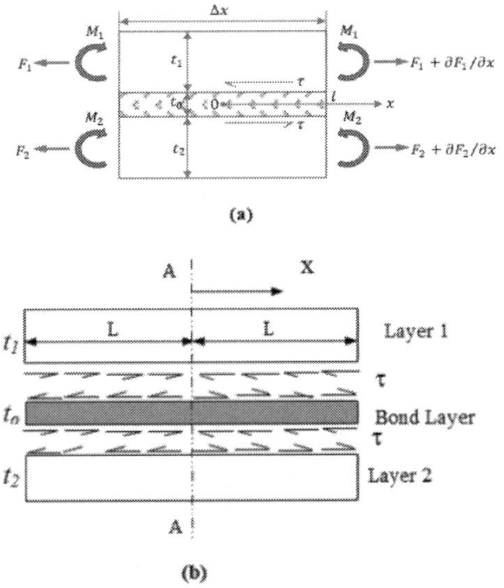

Figure 1. (a) Bi-material assembly and (b) free-body diagram of the model

Symbols and their meanings in this paper:

Material or layer number, i=1, 2; E = Young's modulus (N/m^2); t_i = Thickness (m); ΔT=120°C
α_i = Coefficient of thermal Expansion (1/°C);
ν_i = poison's ratio; R = Radius of curvature;

The key assumptions as follows:

- Each layer can be regarded as Bernoulli beam
- No external force acting among them.
- Axial force due to thermal loading varies along the length of the bonded layers.

The shear stress τ(x) is given by,

$$\tau = \frac{\Delta T(\alpha_1 - \alpha_2)}{K\kappa \, cosh(\kappa L)} sinh(\kappa x) \quad (1)$$

The peeling stress P(x) expression is given by,

$$P = \frac{(t_1 D_2 - t_2 D_1)}{2D} \frac{\Delta T(\alpha_1 - \alpha_2)}{K \, cosh(\kappa L)} cosh(\kappa x) \quad (2)$$

Where $t = t_1 + t_2 + 2t_0$, $\lambda = \lambda_1 + \lambda_2 + \frac{t(t_1 + t_2)}{4D}$,

and $K = K_1 + K_2 + K_0$ and $\kappa = \sqrt{\frac{\lambda}{K}}$

3 Numerical example:

Table 1 shows the parameters used to calculate the interfacial stresses by using the analytical approach. The length of the assembly is $2L = 0.005$ m. The temperature changes, ΔT is taken at 120°C in this computation.

Table 1. Material properties and dimension

Material	Young's Modulus, E_i (GPa)	CTE, α_i (°C), 10^{-6}	Poisson's Ratio	Thickness t_i, mm
Die	110	2.9	0.31	0.40
Die attach	141	22	0.28	0.20
Bond layer	10-70	5-25	0.25-0.33	0.00 - 0.08

4 Results and Discussion

The shearing stress and peeling stress are calculated using eq. (1) and eq. (2) respectively. The results are plotted in Figure 2 to Figure 5 and Table 3 to Table 4.The interfacial stresses are tabulated from x/L = 0.91-1.0, since the stress values are significantly high in the vicinity of the edge. The maximum shearing stress and peeling stress are recorded at the free end ($x/L = 1.0$) as expected.

4.1 Young's Modulus (Stiffness) Effect

Table 2 represents shearing and peeling stresses for different values of Young modulus (E_0) at location x/L = 1. The Young modulus of bond layer, E_0 is varied from 10 to 70 MPa at the interval of 10, 20, 30, 40, and 70 MPa. From the Table 2, it can be observed that the shearing stress is tensile while peeling stress is compressive in nature along the interface as suggested in theory. It is observed that both the shearing and peeling stress increase with the increase of Young's modulus of bond layer. For instance an increase of 60% in shear stress and 83% in peeling stress are observed for an inresae of E_0 from 10 to 70 MPa. Thus, it indicates that a more compliant bond will likely to result in smaller interfacial stress compared to a stiffer bond. However, there might be necessary to compromise between compliant and stiff bond due to the fact that more compliant bond is likely to deform easily [4].

Table2. Shear and peeling stress along the interface with Young's Modulus as parameter

Young's Modulus, E_0 (GPa)	10	20	30	40	70
Shear Stress (MPa)	-43.9	-54.4	-56.0	-63.5	-69.9
Peeling Stress (MPa)	14.5	19.7	22.4	24.1	26.6

4.2 Thickness effect

Table 3 compares shearing and peeling stress with bond layer thickness, t_0 as a parameter at the location x/L= 1

Thickness t_i, mm	0	0.02	0.04	0.06	0.08
Shear Stress (MPa)	-83.2	-77.1	-71.8	-67.3	-63.3
Peeling Stress (MPa)	30.9	27.8	25.2	23.1	21.3

From Table 3, the shearing and peeling stresses decrease with the increase of compliant bond layer thickness. For instance, the shearing and peeling stresses decreased nearly 24% and 31% for an increase of h_0 from 0 to 0.08 mm. The increased bond layer thickness acts as a cushion between the layers which reduces the interfacial shear stress along the interface. Therefore, a higher value of bond layer thickness would be suggested in the flip-chip packaging design. However, the increased bond thickness appears to be another concern for the increased overall space and cost of material.

4.3 Poisson's ratio effect

Table 4 represents the shear and peeling stress values at location x/L = 1 (edge) with Poisson's Ratio varied in the range of 0.25-0.33. It can be observed that the both shear and peeling stress changes very slightly due to the variation of Poisson's Ratio.

Table 4. Stresses at location x/L =1 for different Poisson's ratio

Poisson's Ratio, v	0.25	0.27	0.29	0.31	0.33
Shear Stress (MPa)	-43.9	-43.7	-43.5	-43.2	- 43.0
Peeling Stress (MPa)	29.9	29.8	29.7	29.6	29.5

Thus, Poisson ratio of bond layer do not play significant role in reducing interfacial stresses in layered structure. From the above observation it can be concluded that the effect of Poisson ratio in bond layer may not be essential to consider in predicting stresses development at the interface.

4.4 Thermal expansion coefficient effect

From eq. (1) and Eq. (2), it can be observed that the coefficient of thermal expansion of bond layer term is missing or has no effect in shearing and peeling stress expressions for with bond layer consideration. This is due to the fact that the bond layer is extremely thin compared to the two adhere layers. Therefore, it does not contribute much in terms of expansion and compression of the overall package. Therefore, it can be concluded that the coefficient of thermal expansion of the bond layer is not necessary to consider in predicting stresses development at the interface.

4.5 Temperature effect

From eq. (1) and eq. (2), it can be observed that the bond layer term is also missing in shearing and peeling stress expressions for with bond layer consideration. This is due to the fact that the bond layer is extremely thin compared to the two adhere layers. Therefore, it can be logically assumed that the bond layer temperature will be same as the overall package.

Table 5. Stresses at location x/L =1 for different temperature

Temperature, ℃	60	80	100	120	140
Shear Stress (MPa)	-31.7	-42.3	-52.9	-63.5	-74.1
Peeling Stress (MPa)	17.4	23.2	29.0	34.8	40.6

Nevertheless, Table 5 clearly indicates that the temperature is a very significant factor in stress development in the interface. For instance, shear stress increases 233% for an increment of temperature from 60°C to 120°C at the edge of the package. Therefore, it can be concluded that the temperature should be as low as possible during manufacturing and operation of electronic packages to avoid mechanical and functional failure.

The proposed bond material selection approach:

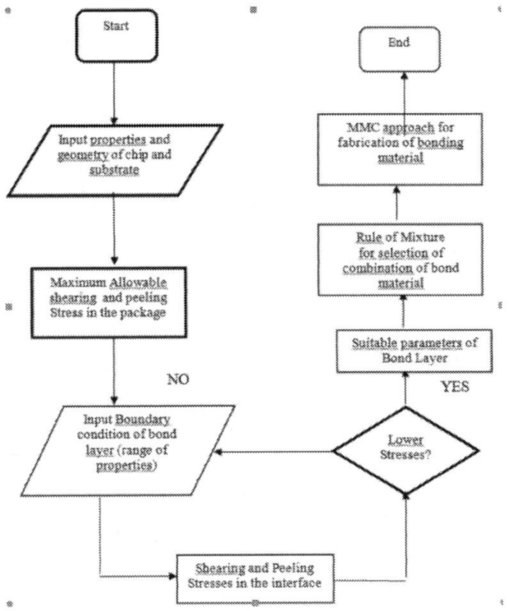

Figure 2: Rule of Mixture design for interfacial bond layer selection

4.6 The detail step-by-step approach for bond material selection, design and fabrication

Step 1: Key in properties and geometry of Die and Die Attach of an arbitrary bi-layered package

Refer to Table 1

Step 2: Key in range of bond layer parameters

Refer to Table 1

Step 3: Key in the interfacial shearing and peeling stress expressions as follows:

The shear stress τ(x) is given by,

$$\tau = \frac{\Delta T(\alpha_1 - \alpha_2)}{K\kappa \, cosh(\kappa L)} sinh(\kappa x) \quad (1)$$

The peeling stress P(x) expression is given by,

$$P = \frac{(t_1 D_2 - t_2 D_1)}{2D} \frac{\Delta T(\alpha_1 - \alpha_2)}{K cosh(\kappa L)} cosh(\kappa x) \quad (2)$$

Step 4: Identify bond layer properties and geometry parameter as follows:

Design data of bond layer for shearing and peeling stresses at $x/L = 1.0$	
*Young's Modulus, E_i (GPa)	50.00
CTE, αi (°C), 10-6	17.25
Poisson's Ratio, ν	0.31
Thickness t_i, mm	0.049

*Only Young's modulus will be considered for developing bond material since other properties are insignificant

Step 5: Find the volume fraction of material combination (alloy) using rule of mixture

The parametric studies which had been carried out earlier concluded that the dominant factors of bond layer in minimizing interfacial stresses in the attached layers are: elastic modulus, E_i and thickness, t_i. Since the thickness of bond layer is a physical property that can be altered, therefore the application of rule of mixture in selecting the material combination for bond layer is focusing on elastic modulus.

The equation for rule of mixture is,

$$E_c = fE_A + (1-f)E_B$$

$f = \frac{V_f}{V_f + V_m}$ (volume fraction) ; *where* E_A = Property of material A and E_B = Property of material B

Table 6. Candidate Lead-free bonding materials in electronic packaging

Material	Young's Modulus, E_i (GPa)
Tin (Sn)	43.0
Silver (Ag)	83
Antimony (sb)	55
Zirconium (Zr)	88
Zinc (Zn)	108
Aluminum (al)	69

 https://doi.org/10.1051/matecconf/201820201005

Gold (Au)	79
Platinum (Pt)	168
Copper (Cu)	117
Indium (In)	11.51
Bismuth (Bi)	34.02
Cadmium (Cd)	62.5

As shown in the Table 6 above, no material from the list exhibits the exact elastic modulus for optimized interfacial stresses in the package, which is 50.0 GPa in Example 1. Therefore, composite or alloy will be a better alternative solution to produce bonding material with desired properties. Examples of alloy/combination used in electronic packaging are: 1. Tin-antimony alloy ; 2. Tin-silver alloy ; 3. Zirconium-tin alloy ; 4. Zinc-aluminium alloy ; 4. Zinc-aluminium alloy ; 5. Gold-platinum alloy ; 7. Silver-copper alloy ; 8. Zinc-Cu-Ti alloy ; 9. Zinc-copper alloy ; 10. Gold-Tin

Example combination 1: Tin-antimony alloy

$$E_C = fE_A + (1-f)E_B$$

E_C = Elastic modulus of desired bonding material, 50 GPa ; E_A = Elastic modulus of tin, 43.0 GPa ; E_B = Elastic modulus of antimony, 55.0 GPa

$$f = \frac{E_C - E_B}{E_A - E_B} = \frac{50.0-55.0}{43.0-55.0} = 0.417$$

Therefore, 41.7% of tin and 59.3% of antimony is required to manufacture tin-antimony alloy bond layer with desired Young modulus, which is 50.0 GPa in order to minimize the interfacial stresses in the silicon-diamond electronic package.

Step 6: Fabrication of MMC composite material based on the combination from Rule of Mixture

5 Conclusions

In this research paper, a methodology is proposed for lead free bond material selection approach. Analytical studies were carried out in order to study the influence of bond layer parameters on interfacial thermal stresses of a given package. From the study, stiffness and bond layer thickness were identified as the key parameters influencing interfacial shearing and peeling stresses. The other parameters namely CTE, poisons ratio has shown insignificant influence on interfacial stresses due to the very thin section of bond layer compared to the top and bottom layers. From the analytical model, it is also obvious that the interfacial stresses increase proportionally with the increase of temperature in the layers. Therefore, temperature should be maintained as low as possible during the chip manufacturing and operating stages. Since only stiffness and bond layer thickness were identified as the key parameters, the interface thermal mismatch stresses can be reduced or eliminated by controlling these two parameters only. Therefore the identification of suitable bond layer parameters selection with reasonable accuracy is possible even without performing optimization process. Finally, this paper proposed a Metal Matrix Composite (MMC) bond material selection approach using rule of mixture material design. The outcome of this research can be seen in the forms of practical and beneficial tools for interfacial stress evaluation and physical design and fabrication of layered assemblies. The Engineers can utilize this research outcome in conjunction with guidelines for electronic packaging under variable thermal properties of layered composites.

Acknowledgement

The authors thankfully acknowledge the research facilities and infrastructure provided by the Curtin University Malaysia and the financial assistance provided by the Ministry of Higher education (Malaysia) under FRGS scheme, Grant reference Code: FRGS/2/2013/SG06/CURTIN/02/1

References

1. Timoshenko. S. (1925). Analysis of Bi-Metal Thermostats. Journal of the Optical Society of America.11, pp. 233-255
2. Tummala, R. (2001). Fundamentals of microsystems packaging . McGraw Hill Professional.
3. Suhir, E. (2017). Flip-chip assembly: Is the bi-material model acceptable? Journal of Materials Science: Materials in Electronics, 28(21), 15775-15781. doi:10.1007/s10854-017-7471-8
4. Moore, G. E. (1998). Cramming more components onto integrated circuits. Proceedings of the IEEE, 86(1), 82-85. doi:10.1109/JPROC.1998.658762
5. Sujan, D., Woldemichael, D. E., Murthy, M. V. V., & Seetharamu, K. N. (2011). Effect of bond layer on bimaterial assembly subjected to uniform temperature change. Journal of Electronic Packaging, Transactions of the ASME, 133(4) doi:10.1115/1.4005294
6. Brown, W. D. (1998). Advanced Electronic Packaging with Emphasis on Multichip Modules. ed.
7. Sujan, D., Pang, X. B., Rahman, M. E., & Reddy, M. M. (2014). Performance of solder bond on thermal mismatch stresses in electronic packaging assembly doi:10.4028/www.scientific.net/MSF.773-774.242
8. Huang, Y.W., Lin, Y. M., Zhan. C.J. Assembly of 3D Chip Stack with 30µm-Pitch Micro Interconnects Using Novel Arrayed-Particles Anisotropic Conductive Film. Proceedings of the 2013 Electronic Components and Technology Conference, pp. 71-76.
9. Suhir, E. (2014). Thermal stress in through-silicon-vias: Theory-of-elasticity approach. Microelectronics Reliability, 54(5), 972-977. doi:10.1016/j.microrel.2014.01.004
10. Gao, L., Chen, X., & Gao, H. (2015). Interfacial thermal stresses in ACF bonding assembly. Microelectronics Reliability, 55(7), 1089-1096. doi:10.1016/j.microrel.2015.04.004

11. Suhir, E. (1989). Interfacial stresses in bimetal thermostats. Journal of Applied Mechanics, Transactions ASME, 56(3), 595-600. doi:10.1115/1.3176133
12. Suhir, E. (1986). CALCULATED THERMALLY INDUCED STRESSES IN ADHESIVELY BONDED AND SOLDERED ASSEMBLIES. Paper presented at the 383-392.
13. Sujan, D., Murthy, M. V. V., Seetharamu, K. N., & Hassan, A. Y. (2005). Complete model for interfacial stresses of a two layered structure. Paper presented at the Proceedings of the 6th International Conference on Thermal, Mechanical and Multi-Physics Simulation and Experiments in Micro-Electronics and Micro-Systems - EuroSimE 2005, , 2005 454-461. doi:10.1109/ESIME.2005.1502848
14. Sujan, D., Murthy, M. V. V., & Seetharamu, K. N. (2011). Improved tri-layered interfacial stress model with the effect of different temperatures in the layers. Archive of Applied Mechanics, 81(5), 561-568. doi:10.1007/s00419-010-0437-4
15. Sujan, D., Murthy, M. V. V., Seetharamu, K. N., & Hassan, A. Y. (2005). Engineering model for interfacial stresses of a heated bimaterial structure with bond material used in electronic packages. Journal of Microelectronics and Electronic Packaging, 2(2), 132-141. doi:10.4071/1551-4897-2.2.132
16. Wu, H. F., Wu, L. L., Slagter, W. J., & Verolme, J. L. (1994). Use of rule of mixtures and metal volume fraction for mechanical property predictions of fibre-reinforced aluminium laminates. Journal of Materials Science, 29(17), 4583-4591. doi:10.1007/BF00376282

Themo-mechanical Interfacial Stress Analysis in Electronic Packaging at Different Temperature Conditions: Revisit Author's Work

D.Sujan, L. Vincent and Y. W. Pok

Faculty of Engineering and Science, Curtin University Malaysia
CDT 250, 98009 Miri, Sarawak, Malaysia

Abstract. The study of thermal mismatch induced stresses and their role in mechanical failure is one relevant topic to composite materials, photonic devices and electronic packages. Therefore, an understanding of the nature of the interfacial stresses under different temperature conditions is necessary in order to minimize or eliminate the risk of mechanical failure. An accurate estimate of thermal stresses in the interfaces plays a significant role in the design and reliability studies of microelectronic devices. In the microelectronic industry, from a practical point of view, there is a need for simple and powerful analytical models to determine interfacial stresses in layered structures. This review paper summarizes the work conducted by the authors in relation to the bi-layered assembly with different temperature conditions on the determination of interfacial thermal stresses. The authors have extended the case of uniform temperature model by earlier researchers of two layered structure to account for differential uniform temperatures, linear temperature gradient in the layers. The presence of a heat source in one layer (die) is also presented. Finally, the effect of bond material properties and geometry on interfacial stresses and bond material selection approach are also considered in a simple way.

1 Introduction

Thermo-mechanical stress develops at the interface of layered structures in electronic packaging during manufacturing (Curing) and operating stages. Since the electronic chips are getting smaller and smaller with increasing demand of power of the devices, a small deviation in the structure will cause functional and mechanical failure of the devices. Therefore, it is very crucial to accurately estimate the interfacial stresses in order to design the devices with reliability [1].

The existing uniform temperature model for bi-material assembly is not really adequate to address the real life situation where the temperature levels of the two layers will be different. Again since there is heat flow in the materials, there will also be a temperature gradient in the layers. Thus, the existence of differential uniform temperatures as well as temperature gradient in the layers should be considered while determining the shearing and peeling stresses at the interface. As a result, a generalized form of the bi-material model is required to be constructed which should be able to take care of any temperature condition in the layers. The effect of heat generation on interfacial stresses due to the presence of a heat source in a layer is also needs to be investigated [1-7].

In this review paper, the authors have presented a summary of work in relation to the bi-layered assembly with different temperature conditions on the determination of interfacial thermal stresses. The authors have extended the case of uniform temperature model by earlier researchers of two layered structure to account for differential uniform temperatures, linear temperature gradient in the layers. The presence of a heat source in one layer (die) is also presented. Finally, the effect of bond material properties and geometry on interfacial stresses and bond material selection approach are also considered in a simple way.

2 Bi-Layered uniform temperature model

Fig.1 shows the full length of the model analyzed. AA represents the centre line of the model. The model length is taken as 2L. In the 2-D model, the model is considered to be of unit width in a direction perpendicular to the plane of the paper and the forces and moments are defined with respect to the unit width.

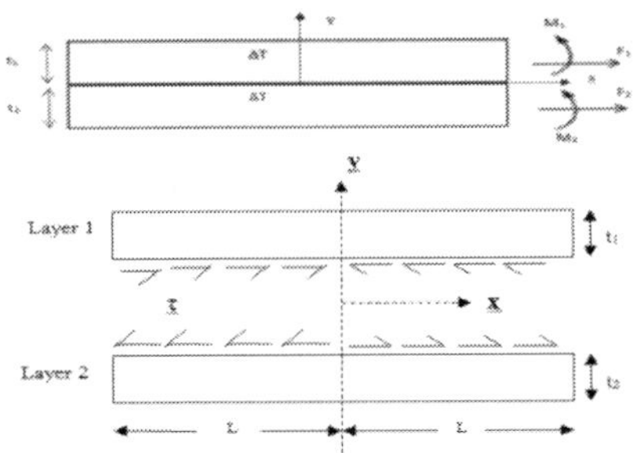

Figure 1: Full Length of the Model [1-2]

The force F at any section of layer is given by

$$F = \int_{-L}^{x} \tau dx , \qquad (1)$$

where τ is the shear stress at the interface

The compatibility condition between top and the bottom layer can be expressed in terms of displacement as:

$$U_{x(1)} - U_{x(2)} = 0 , \qquad (2)$$

where U_i, i=1, 2 are the axial displacements for the layers.

In our approach, we translate the above condition in a more simpler form:

$$\in_{x(1)} = \in_{x(2)} \qquad (3)$$

where $\in_{x(i)}$, i = 1, 2 are the axial strains which is defined as $\in_{x(i)} = \frac{\partial U_i}{\partial x}$

By using condition (3), the model was developed by solving a second order differential equation which is much simpler compared to solution involving integro-differential equation as in the earlier methods.

The solution is based on the assumptions as follows:

1. Thickness of the layered assembly is relatively small.
2. Each layer can be regarded as Bernoulli beam
3. Spherically bending thin plate is acted in each layer.
4. No external force acting among them.
5. Axial force due to thermal loading varies along the length and full shear length in the interface bonded layers.
6. Adhesive layer (solder bond) is very thin compared to the top and bottom layers

With reference to the Figure 1, the axial strain components at the interface of the two layers take the form,

$$\left.\begin{aligned} \in_{x(1)} &= \alpha_1 \Delta T + \lambda_1 F_1 + \frac{t_1}{2R} - K_1 \frac{\partial \tau}{\partial x} \\ \in_{x(2)} &= \alpha_2 \Delta T - \lambda_2 F_2 - \frac{t_2}{2R} + K_2 \frac{\partial \tau}{\partial x} \end{aligned}\right\}, \qquad (4)$$

where, K_i = interfacial shear compliances given by $K_i = \frac{h_i}{3G_i}$; G_i = shear modulus of rigidities given by $G_i = \frac{E_i}{2(1+\nu_i)}$; λ_i , *i* = Axial Compliances which is given by $\lambda_i = \frac{(1-\nu_i^2)}{E_i h_i}$

The shear strain components in the layers in equation (4) are expressed as follows:

(1) Strain due to shearing force = $\pm K_i \frac{\partial \tau}{\partial x}$;

(2) Strain due to axial force $F_i = \pm \lambda_i F_i$;

(3) Strain due to change of curvature = $\pm \frac{h_i}{2R}$;

(4) Strain due to change of temperature = $+\alpha_i \Delta T$ (+ve sign because ΔT is assumed to be positive and consequently the effect is an extension in the layers)

The shear stress τ(x) is given by,

$$\tau = \frac{\Delta T(\alpha_1 - \alpha_2)}{K\kappa \, cosh(\kappa L)} sinh(\kappa x) \qquad (5)$$

The peeling stress P(x) expression is given by,

$$P = \frac{(t_1 D_2 - t_2 D_1)}{2D} \frac{\Delta T(\alpha_1 - \alpha_2)}{K \, cosh(\kappa L)} cosh(\kappa x) \qquad (6)$$

In eq (5), and (6), $\kappa^2 = \frac{\lambda}{K}$ and $K = K_1 + K_2$

where $K_1 = \frac{t_1}{3G_1}$ and $K_2 = \frac{t_2}{3G_2}$ are shear stress compliances for upper and lower layer respectively.

2.1 Bi-Material differential temperatures model

https://doi.org/10.1051/matecconf/201820201006

Introducing two parameters relating to temperature and thermal expansion coefficient namely $m = \dfrac{\Delta T_2}{\Delta T_1}$ and $n = \dfrac{\alpha_2}{\alpha_1}$, equations (5) and (6) can be expressed as,

$$\tau(x) = \frac{\alpha_1 \Delta T_1 (1 - mn)}{K\kappa \, cosh(\kappa L)} sinh(\kappa x), \tag{7}$$

$$P(x) = \frac{\left(t_1 D_2 - t_2 D_1\right)}{2D} \frac{\alpha_1 \Delta T_1 (1 - mn)}{K cosh(\kappa L)} cosh(\kappa x) \tag{8}$$

Where temperature changes assigned for entire Layer 1 = ΔT_1 and for entire Layer 2 = ΔT_2.

2.2 Bi-Material Linear Temperature Gradient model

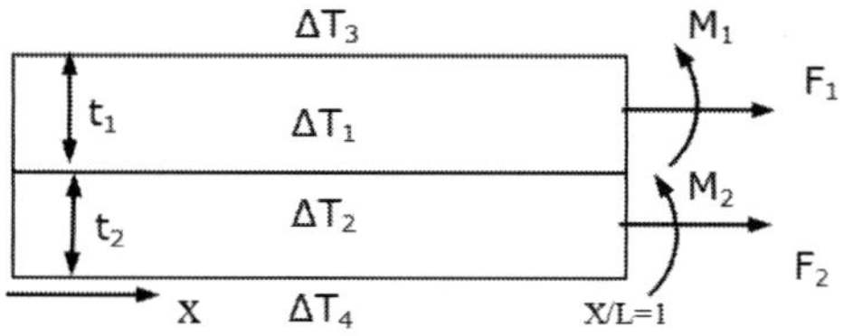

Figure 2: Bi-layered Assembly with Linear Temperature Gradients in the Layers [1].

Considering the top layer of Fig. 2, the temperature distribution throughout the thickness can be represented as shown in Fig 3.

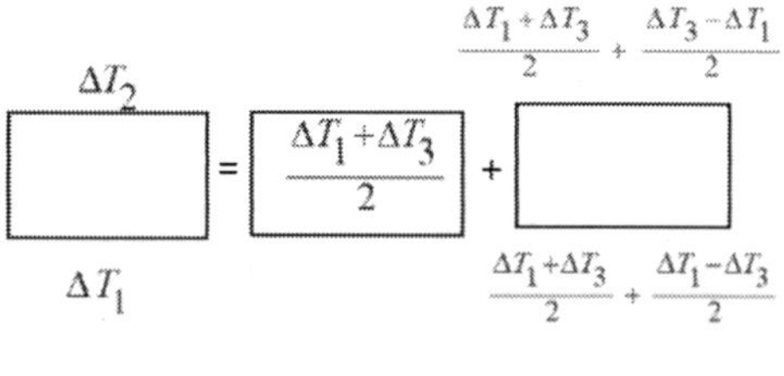

Figure 3: Linear Temperature Distribution Gradient in the Top Layer [1].

The total change of curvature of the assembly due to change of temperature is expressed by

$$\frac{1}{R} = \frac{1}{R_{1(T)}}\left(\frac{D_1}{D}\right) + \frac{1}{R_{2(T)}}\left(\frac{D_2}{D}\right) + \frac{hF}{2D} \tag{9}$$

Now considering this modified value of $\frac{1}{R}$ in eq. (9), the eq. (7) and (8) can be reconstructed as follows:

$$\tau = \frac{\alpha_1 \Delta T_1 (A_1 + A_2)}{K \mu cosh \mu L} sinh \mu x \tag{10}$$

$$P = \frac{\left(h_1 D_2 - h_2 D_1\right)}{2D} \frac{\alpha_1 \Delta T_1 (A_1 + A_2)}{K \cosh \mu L} \cosh \mu x, \tag{11}$$

where $A_2 = \beta_1 \gamma_1 - mn\beta_2 \gamma_2$, (12)

where $\beta_1 = \dfrac{\Delta T_1 - \Delta T_3}{\Delta T_1}$, $\beta_2 = \dfrac{\Delta T_2 - \Delta T_4}{\Delta T_2}$,

$\gamma_1 = \dfrac{tD_1}{2t_1 D}$ and $\gamma_2 = \dfrac{tD_2}{2t_2 D}$.

From eq. (12) it can be observed that when gradient in materials is zero ($\Delta T_1 = \Delta T_3$) and ($\Delta T_2 = \Delta T_4$), the term A_2 becomes zero and eq. (10) and (11) reduces to (7) and (8), which are the differential uniform temperature model. It is also observed that the correction factor A_2 (in eq. 12) is contributed by six parameters m, n, β_1, β_2, γ_1, and γ_2.

2.3 Derivation for heat generation in the layer (Die)

In reality heat is generated in one of the layers say die. In this situation, the temperature distribution gradient accross the layer is expected to be quadratic ruther than linear which is expressed by the differential equation,

$$\frac{d^2 T}{dy^2} + \frac{G'}{k'} = 0 \quad , \tag{13}$$

where G' and k' represent heat flux and thermal conductivity of the die material respectively [10].

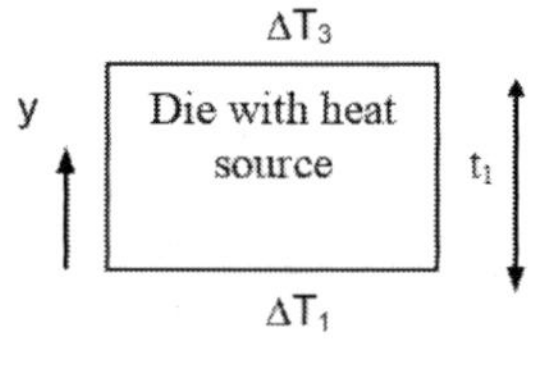

Figure 4. A Die Section with a Heat Source [1]

The expression of the ΔT term can be formed by applying boundary conditions in Fig. 4, at y=0, $\Delta T = \Delta T_1$ and at $y = t_1$, $\Delta T = \Delta T_3$, the solution of eq. (13) is given by,

$$\Delta T = \frac{G'}{2k'}(t_1 - y)y - \frac{(\Delta T_1 - \Delta T_3)}{t_1} y + \Delta T_1, \tag{14}$$

where ΔT_1 and ΔT_3 represent temperature changes at the interface and top of the die respectively. Thus, so far the analytical model has taken account of the differntail temperature conditions in the layers which is more realistic from the practical packaging point of view.

3 Analytical Model with Bond layer

In the previous sections, a perfect bonding condition was assumed in the development of the interfacial stress model for bi-layered electronic package. However in reality, there exists a very thin layer of adhesive bonding material to attach the two layers together. There there is a need to evaluate the influence of this tiny bond layer in the analytical model. Interestingly, this surrogate bond layer may contribute significantly in eliviating the interfacial stresses by choosing appropriate bond layer parameters. In this section, the previous bi-layered perfect model in section 1 is further upgraded with the bond layer consideration. Subsequently a process flow chart is proposed to select the suitable bond using rule of mixture material for physical design and fabrication of layered assemblies.

The same analytical model which has been used in paper 1 (Title: Bond layer properties and geometry effect on interfacial thermo-mechanical stresses in bi-material electronic packaging assembly) in this conference is utilized for bond material selection, and design approach. In this paper, only the final model is presented. Figure 5 shows the free body diagram of the full length of the model.

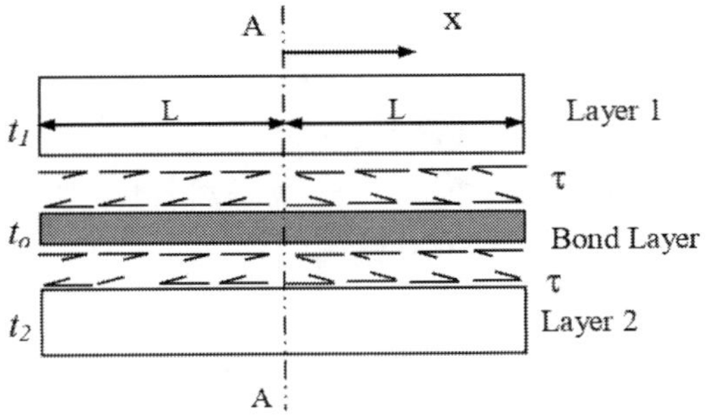

Figure 5. free-body diagram of the model [3-4, 7]

The shear stress τ(x) is given by,

$$\tau = \frac{\Delta T(\alpha_1 - \alpha_2)}{K\kappa\ cosh(\kappa L)} sinh(\kappa x) \qquad (15)$$

The peeling stress P(x) expression is given by,

$$P = \frac{(h_1 D_2 - h_2 D_1)}{2D} \frac{\Delta T(\alpha_1 - \alpha_2)}{K\,cosh(\kappa L)} cosh(\kappa x) \quad (16)$$

Where $t = t_1 + t_2 + 2t_0$, $\lambda = \lambda_1 + \lambda_2 + \frac{t(t_1 + t_2)}{4D}$,

and $K = K_1 + K_2 + K_0$

Shear compliance for the bond layer, $K_0 = \frac{t_0}{G_0}$

4 The proposed bond material selection approach

Step 1: Key in properties and geometry of chip and substrate of an arbitrary bi-layered package

Step 2: Key in range of bond layer parameters

Step 3: Key in the interfacial shearing and peeling stress

Step 4: Bond layer property and geometry parameter (for instance)

Design data of bond layer for shearing stress and peeling stress at $x/L = 1.0$

Young's Modulus, E_l (GPa)	50.00
CTE, αi (°C), 10-6	17.25
Poisson's Ratio, ν	0.31
Thickness h_l, mm	0.049

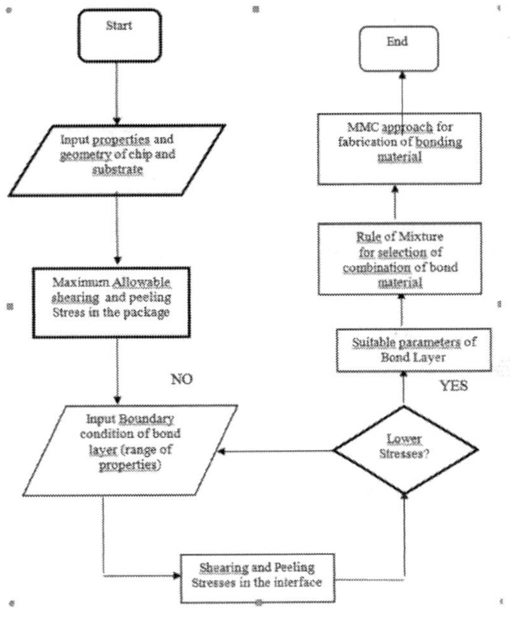

Figure 6: Rule of Mixture design for interfacial bond layer selection

Step 5: Find the volume fraction of material combination (alloy) using rule of mixture

The parametric study carried out earlier concluded that the dominant factors of bond layer in minimizing interfacial stresses in the attached layers are: elastic modulus, E_l and thickness, t_l. Since the thickness of bond layer is a physical property that can be altered, therefore the application of rule of mixture in selecting the material combination for bond layer is focusing on elastic modulus.

The equation for rule of mixture is,

$$E_c = fE_A + (1 - f)E_B$$

$f = \dfrac{V_f}{V_f + V_m}$ (volume fraction) ; *where* E_A = Property of material A and E_B = Property of material A

Example combination 1: Tin-antimony alloy

$$E_c = fE_A + (1 - f)E_B$$

E_C = Elastic modulus of desired bonding material, 50 GPa ; E_A = Elastic modulus of tin, 43.0 GPa ; E_B = Elastic modulus of antimony, 55.0 GPa

$$f = \frac{E_C - E_B}{E_A - E_B} = \frac{50.0 - 55.0}{43.0 - 55.0} = 0.417$$

Therefore, 41.7% of tin and 59.3% of antimony is required to manufacture tin-antimony alloy bond layer with desired Young modulus, which is 50.0 GPa in order to minimize the interfacial stresses in the silicon-diamond electronic package.

Step 6: Fabrication of MMC composite material based on the combination received from Rule of Mixture

5 Conclusions

This review paper summarizes the work conducted by the authors in relation to the bi-layered assembly with different temperature conditions on the determination of interfacial thermal stresses. The authors have extended the case of uniform temperature model by earlier researchers of two layered structure to account for differential uniform temperatures, linear temperature gradient in the layers. The presence of a heat source in one layer (die) is also presented. Finally, the effect of bond material properties and geometry on interfacial stresses and bond material selection approach are also considered in a simple way.

Acknowledgement

The authors thankfully acknowledge the research facilities and infrastructure provided by the Curtin University Malaysia and the financial assistance provided by the Ministry of Higher education (Malaysia) under FRGS scheme, Grant reference Code: FRGS/2/2013/SG06/CURTIN/02/1

References

1. Sujan, D., Murthy, M. V. V., Seetharamu, K. N., & Hassan, A. Y. (2005). Complete model for interfacial stresses of a two layered structure. Paper presented at the Proceedings of the 6th International Conference on Thermal, Mechanical and Multi-Physics Simulation and Experiments in Micro-Electronics and Micro-Systems - EuroSimE 2005, , 2005 454-461. doi:10.1109/ESIME.2005.1502848
2. Debnath, S., Rahman, M. E., Engida, W. D., Murthy, M. V. V., & Seetharamu, K. N. (2011). Bi-layered model of interfacial thermal stresses with the effect of different temperatures in the layers doi:10.4028/www.scientific.net/AMM.87.63
3. Sujan, D., Woldemichael, D. E., Murthy, M. V. V., & Seetharamu, K. N. (2011). Effect of bond layer on bimaterial assembly subjected to uniform temperature change. Journal of Electronic Packaging, Transactions of the ASME, 133(4) doi:10.1115/1.4005294
4. Sujan, D., Murthy, M. V. V., Seetharamu, K. N., & Hassan, A. Y. (2005). Engineering model for interfacial stresses of a heated bimaterial structure with bond material used in electronic packages. Journal of Microelectronics and Electronic Packaging, 2(2), 132-141. doi:10.4071/1551-4897-2.2.132
5. Brown, W. D. (1998). Advanced Electronic Packaging with Emphasis on Multichip Modules. ed.
6. Suhir, E. (1989). Interfacial stresses in bimetal thermostats. Journal of Applied Mechanics, Transactions ASME, 56(3), 595-600. doi:10.1115/1.3176133
7. Sujan, D., Pang, X. B., Rahman, M. E., & Reddy, M. M. (2014). Performance of solder bond on thermal mismatch stresses in electronic packaging assembly doi:10.4028/www.scientific.net/MSF.773-774.242

Design Optimization of Interior Double-Radial Synthetic Magnetic Field Permanent Magnet Generator for Electric Vehicle

Shilun Ma ,Xueyi Zhang and Wenjin Hu

School of Transpotation and Vehicle Engineering,Shandong University of Technology,255049,ZiBo,China

Abstract. The Interior double-radial permanent magnent generator (IDRPMG) which composed by two groups of rectangular permanent magnets to provide parallel magnetic circuits of the rator and the sator core with less eddy current loss, low hormonic content and low cogging torque of the stator with fractional slot winding is developed. It has the advantages of remarkable magnetism gathering effect, strong magnetic field intensity and high space utilization.Combining Taguchi method and finite element method, the relevant parameters of the permanent magnet size and the angle between the first and second rectangle permanent magnets in rotor are optimized to get better the distortion rate of output voltage waveform, lower cogging torque and higer peak value of airgap flux density. Then finite element simulation is taken for the best optimization scheme through comparative analysis of the machine by before and after optimization.It showed that each performance index is improved after optimization. Finally, the prototype is manufactured, according to the optimization parameters and some experiments are conducted, which results verify the analys is preview well.

1 Introduction

Due to the permanent magnet generator has no excitation winding, carbon brush slip ring and brush compared to the silicon rectifying generator, it has the advantages of low loss, high efficiency and simple structure[1-2], So it's widely used in electric vehicle. However, the permanent magnet generator also has the inevitable defects such as the large cogging torque, high harmonic content and large magnetic flux leakage which will affect the output characteristics of permanent magnet generator[3-5]. In the literature[6], a new surface-Mounted permanent magnet generator is presented which the magnetic poles consists of a magnetic metal block and a permanent magnet. The generator uses the non-uniform air gap structure to reduce air gap flux density harmonics content and improve the output voltage waveform, but the disadvantage of this structure is that large magnetic flux leakage and low power generation efficiency. In the literature [7] by using the multi-layer permanent magnet instead of the traditional single layer permanent magnet as the magnetic pole of the generator to improve the air-gap magnetic waveform, but this structure requires more permanent magnet so that the cost is high. In the literature [8], by optimizing the shape of the permanent magnet with unequal thickness, the harmonic39 content and cogging torque are reduced, but at the same time, increasing machining difficulty of permanent magnet. It is necessary to develop a permanent magnet generator which has less power consumption, higher power generation efficiency, lower cogging torque and steady output voltage.

2 Multi-objective optimization design method of IDRPMG

2.1 Initial design parameters of IDRPMG

The initial design parameters of permanent magnet generator with double-radial synthetic magnetic field are obtained through the calculation of empirical formula. The results are shown in Table 1.

Table 1. Intial design parameters of IDRPMG.

Parameters	Values	Parameters	Values
The rated voltage(V)	14	Width of the first rectangular permanent magnet(mm)	11
The rated power(W)	1000	Width of the second rectangular permanent magnet(mm)	5
The pole pair number	4	Thickness of the first rectangle permanent magnet(mm)	3.5
Number of stator slots	36	Thickness of the second rectangle permanent magnet(mm)	2.5

Inner diameter of stator(mm)	91	The angle between the first and second permanent magnets(°)	46
Outer diameter of stator(mm)	126	The Outer diameter of rotor(mm)	90
The shape of the stator slot	Pyriform	The axial length(mm)	52

2.2 Design methods and objectives

In this paper, the multi-objective optimization design of permanent magnet generator is realized by Taguchi method to determine the experimental condition and calculate the target characteristics under specific test conditions by establishing orthogonal table,this method can obtain the best combination of parameters in the least number of experiments, thus realizing the optimization design of multi-objective characteristics[9-10]. In this study, the width of the first rectangular permanent magnet (b_1), magnetization direction thickness of the first rectangle permanent magnet (h_1), the width of the second rectangular permanent magnet(b_2), magnetization direction thickness of the second rectangle permanent magnet (h_2) and the angle between the first and second permanent magnets (θ) is selected as the optimization of parameters. As shown in Figure 1.

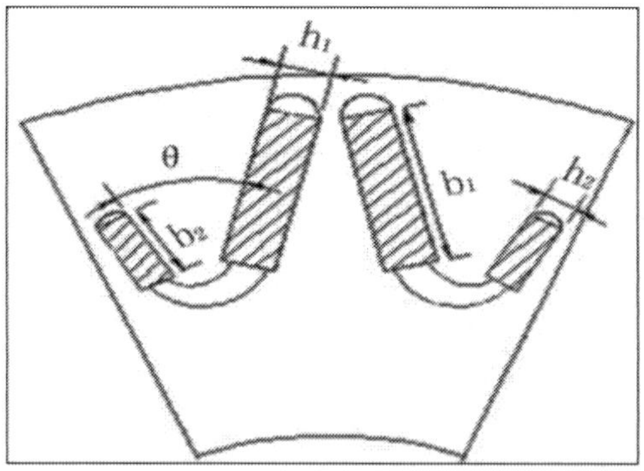

Figure 1. The schematic diagram of adjacent N and S poles

Each parameter selects 4 levels of influence factors,and the levels influence factors of parameters are selected near the initial design parameters.The cogging torque peak (T),the voltage waveform distortion rate (Kr) and the air gap flux density peak (G) are selected as the optimization objectives,the influence factor level shown in Table 2.

Table 2. The influence factors of structural parameters of permanent of permanent magnet generator

Level Factors	b_1(mm)	h_1(mm)	b_2(mm)	h_2(mm)	θ(°)
1	10	3.5	6	2	50
2	11	3	6.5	2.5	46
3	12	2.5	5.5	3.3	42
4	13	2	7	3	48

The calculation formula of output votage waveform distortion rate of permanent magnet generator:

$$K_\rho = \frac{\left(Y_2^2 + Y_3^2 \cdots\cdots + Y_N^2\right)^{\frac{1}{2}}}{Y_1} \times 100\% \quad (1)$$

Where Y_N is the amplitude of Nth harmonic.

2.3 Optimization process

The orthogonal experiment selects 5 parameter variables, and each parameter has four level factor.Thus,the experimental matrix of expression $L_{16}(4^5)$ is established, If the traditional form of single variable and single objective optimization method are used, 4^5=1024 experiments are required. However, the Taguchi method can be used to optimize the multi-parameter and multi-objective design of the permanent magnet generator with only 16 times experiments.The experimental orthogonal table and simulation results are shown in Table 3.

Table 3. Experimental matrix and the results of finite element analysis

NO.	Experimental matrix					Test evalution index		
	b_1	h_1	b_2	h_2	θ	K_r (%)	T (N·m)	G (mT)
1	1	1	1	1	1	15.16	0.069	732
2	1	2	2	2	2	16.63	0.261	719
3	1	3	3	3	3	16.40	0.212	600
4	1	4	4	4	4	17.10	0.269	675
5	2	1	2	3	4	16.25	0.105	783
6	2	2	1	4	3	14.70	0.227	865
7	2	3	4	1	2	17.00	0.291	811
8	2	4	3	2	1	17.13	0.230	609
9	3	1	3	4	2	16.39	0.082	828
10	3	2	4	3	1	17.83	0.205	830
11	3	3	1	2	4	18.20	0.231	745
12	3	4	2	1	3	17.41	0.304	644
13	4	1	4	2	3	15.88	0.154	846
14	4	2	3	1	4	17.47	0.139	916
15	4	3	2	4	1	18.41	0.244	860
16	4	4	1	3	2	18.06	0.212	765

The average value is calculated in the solution of Table 3,the calculation formula as follows:

$$m = \frac{1}{n}\sum_{i}^{n} S_i \quad (2)$$

Where n is experiment times. S_i is the value of the target performance of the ith experiment. The results are shown in Table 4.

Table 4. The average value of target performance indicators

Optimization	K_r(%)	T(N·m)	G(mT)
Index	16.88	0.202	764.5

Then the average value of each factor of the parameter is calculated for the specific target performance, the calculation formula as follows:

$$m_{xi} = \frac{1}{4}[m_x(j) + m_x(k) + m_i(l) + m_x(n)] \quad (3)$$

Where m_{xi} is the average value of the performance index of the first influnce factor of perameter X; m_x is the performance index of parameter *X*. m_x is the performance index of parameter x in one experiment; *J, k, l, n* is Serial number of test.

In order to facilitate the comprehensive analysis, the index is represented by the line graph with the level change of factors,as shown in Figure 2.

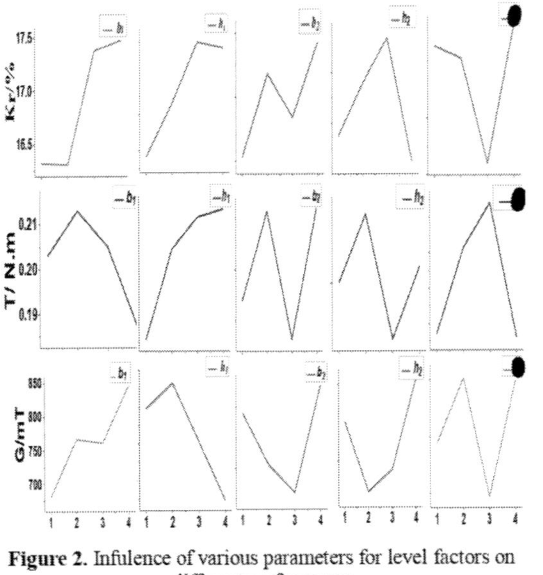

Figure 2. Infulence of various parameters for level factors on different performance

We can see from Figure 2, with the increase of the width of the first rectangle permanent magnet b_1, the decrease of the thickness of magnetization direction h1 and the decrease of angle θ, the distortion rate of the voltage waveform of the permanent magnet generator becomes larger.While with the decrease of the thickness of the first rectangular permanent magnet h_1, the opening angle between the first and second rectangular permanent magnent θ or the increase of the width the second rectangular permanent magnet, the peak value of the cogging torque is increased. In addition, when the width and thickness of the first rectangular permanent magnet are increased, the peak value of the air gap flux density of the permanent magnet generator is also increased. From the Figure 5,we can obtained the combination of the parameters which the smallest distortion rate of the voltage waveform is $b_1(2)h_1(1)b_2(1)h_2(1)\theta(3)$, the combination of the parameters which the minimum peak value of cogging torque is $b_1(4)h_1(1)b_2(3)h_2(3)\theta(1)$,the combination of the parameters which the maximum value of the peak value of air gap flux density is $b_1(1)h_1(2)b_2(4)h_2(4)\theta(2)$. Obviously, the three groups of level combinations are designed for the optimization of a single performance index, if considering the influence of three performance indexes on the permanent magnet generator, it is necessary to analyze the variance to further analyze the influence of the change of each parameter on the different performance indicators, and thus obtain the optimization results.

2.4 The analysis of proportion of each parameter to permance index

Analyze the variance of the average value of a performance index under different parameters at different levels for all the experimental performance indexes,We can determine which parameters had significant effect on the performance index, the results are shown in Table 5.The formula for calculation is as follows:

$$S_s = \frac{1}{Z}\sum_{i=1}^{z}(Q_{(i)} - Q)^2 \qquad (4)$$

Where s is the influence factor, such as b_1,h_1,b_2,h_2,θ; Ss is the variance of a performance index under parameter,in this study, Z=4; $Q_{(i)}$ is the average value of the parameters in the level of factor i under certain performance index; Q is the total average value of permance index.

Table 5. The variance and proportion of performance index of each parameter.

Parameter	S_{kr}		S_T		S_G	
	Variance	Proportion(%)	Variance	Proportion(%)	Variance	Proportion(%)
b_1	0.340	37.32	9.2×10^{-5}	1.67	3427	43.14
h_1	0.410	45.02	3.7×10^{-3}	67.40	3558	44.79
b_2	0.055	6.03	7.6×10^{-4}	13.85	421	5.30
h_2	0.034	3.73	6.7×10^{-4}	12.22	236	2.97
θ	0.072	7.90	2.67×10^{-4}	4.86	302	3.80
Total	0.911	100	5.49×10^{-3}	100	7944	100

2.5 Determination of final optimizition scheme

From Table 5 we can found that the value of variance can directly reflect the influence of each optimization parameter on the performance index[11]. The thickness of the the first rectangular permanent magnet has the great influnce on three optimation indexes. Compared with other optimization indexs, the parameter b_1 has the significant influence on the peak value of air gap flux density.Among the three optimation indexes, the change of the opening angle between the first and the second rectangular permanent magnet has the largest influence on the voltage waveform distortion rate.

According to the above analysis, we selected the parameter b_1 and h_1, according to the optimization standards which the minimum distortion rate of the voltage waveform and the maximum peak value of the air gap flux density as the optimization standards, and the parameter θ is chosen as the optimization criterion of the minimum distortion rate of the voltage waveform.The final optimization result is $b_1(4)h_1(1)b_2(1)h_2(1)\theta(3)$.

3 Finite element analysis simulation analysis of IDRPM

According to the parameter values of the final scheme, the two-dimensional of double-radial synthetic magnetic field of permanent magnet generator is built, and the finite element simulation analysis is carried out.The simulation results are shown in Figures 3 and 4. In Figure 3,The main magnetic flux of permanent magnet generator is supplied by the first and second rectangular permanent magnets. The magnetic flux provided by the first rectangular permanent magnet is parallel with the magnetic flux provided by second rectangular permanent magnet ,the magnetic field distribution is uniform and the magnetic flux is less. In Figure4, the different colours represent different magnetic density in magnetic flux desity modulus maps,the magnetic density values of each part of permanent magnet generator can be obtained by the color distribution, the magnetic flux density distribution is uniform and the stator yoke flux density is 1.8 T,the simulation results meet the design requirements.

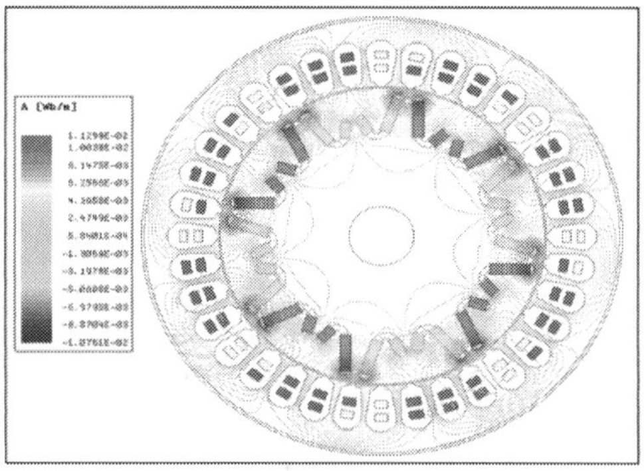

Figure3. The distribution of magnetic induction lines

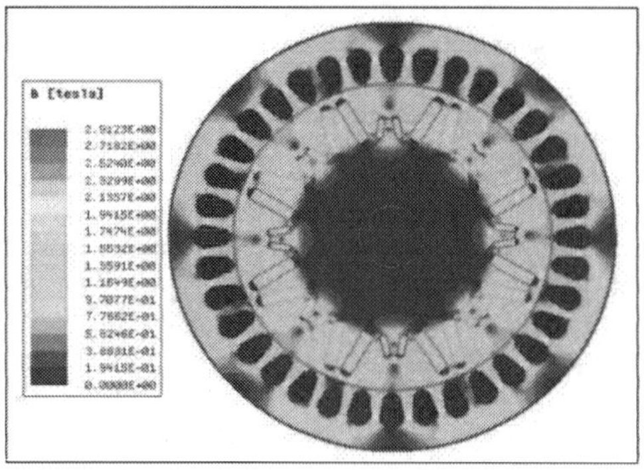

Figure4. The modulus values of magnetic flux density

In Figure 5 ,we can see that the peak value of cogging torque of permanent magnet is 0.23N·m before optimization, while the peak value of cogging torque reduced to 0.043N·m after optimization, the results of finite element simulation analysis shows that the cogging torque of the permanent magnet generator with double radial synthetic magnetic field is greatly weakened and the rationality of the design is verified.

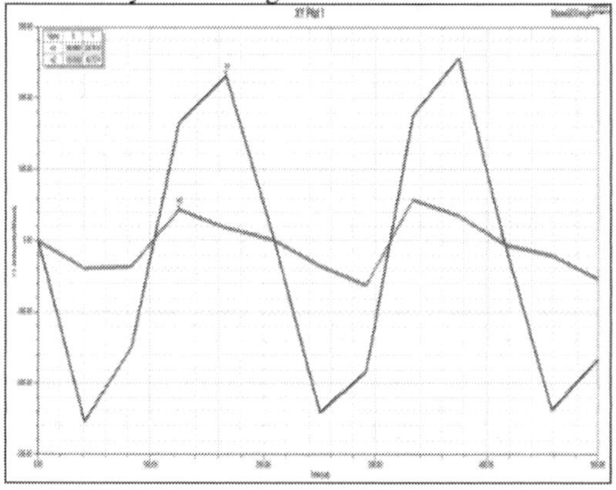

Figure 5. The comparison of cogging torque waveform before and after optimization

In Figure 6, we can see that the output voltage fluctuates greatly and the top of the voltage waveform shows obvious depression before optimization.In Figure7, the voltage waveform has a small fluctuation after optimization,but compared to before optimization more close to the sine wave.After optimization,the amplitude of the output voltage of the permanent magnet generator is reduced,the distortion rate of the voltage waveform is 15.6%,which is 1.6% lower than before optimization.In Figure 9, it can be seen that the peak value of air gap magnetic density is 705mT before optimization, after optimization the peak value of air gap magnetic density is 840mT, which is up about 20%.

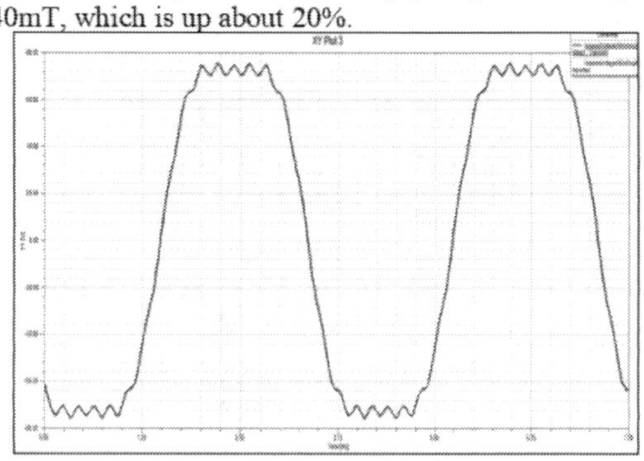

Figure 6. The Output voltage of permanent magnet generator before optimazition

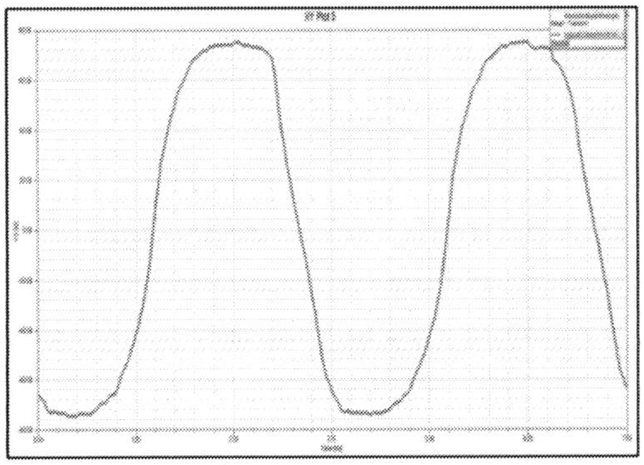

Figure 7. The Output voltage of permanent magnet generator after optimazition

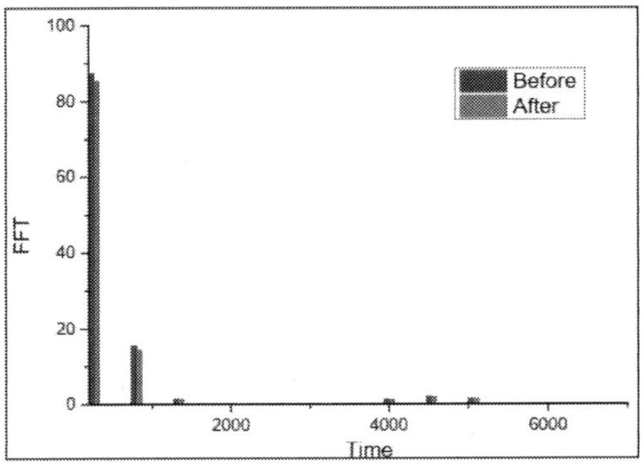

Figure 8. The harmonic amplitude of output voltage before and after optimization

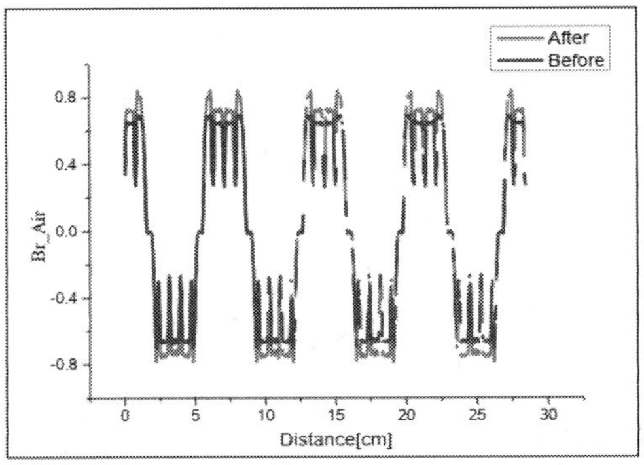

Figure 9. The peak value of air gap flux densitybefore and after optimization

4 Conclusion

(1) In this study, a new type of built in double radial synthetic magnetic field permanent magnet generator is developed, the generator is composed of two groups of rectangularpermanent magnets to produce parallel magnetic circuit, and the resulting magnetic fiedis synthesized in the air gap. The shunt circuit magnetic path can form a significant magnet magnet congregate effect, the air gap has large flux density and high efficiency.

(2) Compared Taguchi method and finite element method to optimize the design of permanent magnet generator, after optimized the output voltage distortion rate decreased 1.6%, the peak value of cogging torque is weakened by 81.3%, the air gap flux density peak is increased about 20%.

References

1. X. Zhang, Q.Du, S.Ma, H.Geng, W.Hu, Z.Li. IEEE Access, *Permeance Analysis and Calculation of the Double-Radial Rare-Earth Permanent Magnet Voltage-Stabilizing Generation Device*,99,1-1(2018).
2. Zhang XY, Shi LW, Wang YJ. Int J Electr Hybrid Veh, *Design and stabilivolt analysis of Nd-Fe-B permanent magnet generator for electric vehicle range extende*. **3**,259-271 (2011).
3. Aimeng Wang,Wenyuan Xi, Hui Wang.Electr Mach Contrl.*Optimization Design of Interior permanent Magnet Synchronous Motor Based on Hybrid Genetic Algorithm*.**03**,59-65+95 (2017).
4. X.Cao, G.Li, Q.Ye, R.Zhou, G.Ma, F. Zhou. ICIEA, *Multi-objective optimization of permanent magnet synchronous motor based on elite retention hybrid simulated annealing algorithm*,535-540 (2017).
5. Si Jikai, Zhang Lufeng, Feng Haichao, Xu Xiaozhuo. J CHINA COAL SOC. *Multi-objective optimal design of a surface-mounted and interior permanent magnet synchronous motor*.**12**,3167-3173(2016).
6. Zhang BingYi, Jia Yiqi, LiKai Feng. Guihong.Electr Mach Contrl. *Study on magnetic pole structure of surface mounted PMSM*. **05**,43-48(2014).
7. D. K. Hong, W. H.Wang, J. Y. Lee, B. C. Woo.IEEE Trans Magn. *Design, Analysis, and Experimental Validation of a Permanent Magnet Synchronous Motor for Articulated Robot Applications*.**54**,1-4(2018).
8. Si Jikai, Zhang Lufeng, Feng Haichao, Xu Xiaozhuo.J CHINA COAL SOC. *Characteristic analysis of surface-mounted and interior hybrid permanent magnet synchronous motor equivalent magnetic circuit method*. **05**,1199-1205(2015).
9. Xueyi Zhang, Qinjun Du, Chao Ma, Shilun Ma, Jinbin Xu ,Huihui Geng, Guangdong Tian. ADV MECH.*Nd-Fe-B permanent magnet generator and voltage stabilizing control technology for vehicles*. **8(9)**,1-11(2017).
10. Huijun Wang, Zheng Qu, Shaofei Tang, JMMM,Mingqi Pang, Mingju Zhang, *Analysis and optimization of hybrid excitation permanent magnet synchronous generator for stand-alone power system*. **436**, 117-125(2017).
11. Xia C, Guo L, Zhang Z. IEEE Trans Magn. *Optimal Designing of Permanent Magnet Cavity to Reduce Iron Loss of Interior Permanent Magnet Machine*. , **51(12)**:1-9(2015).

Theoretical Analysis on the Impact of Total Damping Ratio on the Power Output of an Electromagnetic Vibration Energy Harvester

Faruq Muhammad Foong[1], Chung Ket Thein[1] and Beng Lee Ooi[2]

[1]*School of Engineering and Physical Sciences, Heriot Watt University Malaysia*
[2]*Intel PSG, PG 14, Plot 6, Bayan Lepas Technoplex, Medan Bayan Lepas, 11900 Penang, Malaysia.*

Abstract. Vibration energy harvesting has emerged as a promising source of sustainable energy to power small electronics. This study investigates the effect of total damping on the power output of an electromagnetic vibration energy harvester. Analytical results show that an increase in the effective mass of the harvester increases the mechanical damping but decreases the electromagnetic damping. The total damping of the harvester displayed an increasing trend with the effective mass when the electromagnetic damping is lower that the mechanical damping but changed into a decreasing trend when the electromagnetic damping becomes larger than the mechanical damping. Findings also suggest that there is an optimum proof mass to beam mass ratio where the harvester would produce maximum power in both cases of where a constant and varying optimum load resistance were considered.

1 Introduction

Advancement in wireless sensor network (WSN) technologies have increased research in energy harvesting to find a sustainable source of energy to power these devices. While sources such as heat, light and wind energy harvesting was an option, vibration energy harvesting became an instant research attraction due to its promising advantages [1–3].

There exists several methods to convert the mechanical vibrational energy into electrical power with the two most common methods being piezoelectric conversion and electromagnetic induction. Kim et al. [4] reported that piezoelectric harvesters perform better at small volumes whereas electromagnetic harvesters surpassed piezoelectric at larger volumes. The performance of an electromagnetic vibration energy harvester is strongly dependent on the velocity of the vibrating beam, which can be strongly related to the damping of the system.

Generally, an electromagnetic harvester consist of two main sources of damping, which are the mechanical damping and the electromagnetic damping [5]. For a cantilever-beam based harvester, the mechanical damping originates from the internal friction of the vibrating beam. Damping due to internal friction is highly material specific. Hence, different material will exhibit different damping capacities. In an earlier study, Lazan [6] proposed that the mechanical damping of a structure can be strongly related to its maximum vibrating stress and its fatigue limit stress. The study suggest an increase in mechanical damping when the stress of the structure increase, which is generally true.

Electromagnetic damping arises from the interaction between the eddy current in a vibrating coil and a magnetic field, producing a force that retards the vibrating motion [7]. Consequently, the total damping of an electromagnetic harvester is simply the sum of its mechanical and electromagnetic damping. Damping plays a complex role in influencing the power output of an electromagnetic energy harvester, and this study aims to investigate the effect of the total damping on the power output of the harvester through analytical means.

2 Governing equations

Figure 1 shows the mechanism of a typical cantilever beam-based electromagnetic vibration energy harvester under harmonic base excitation.

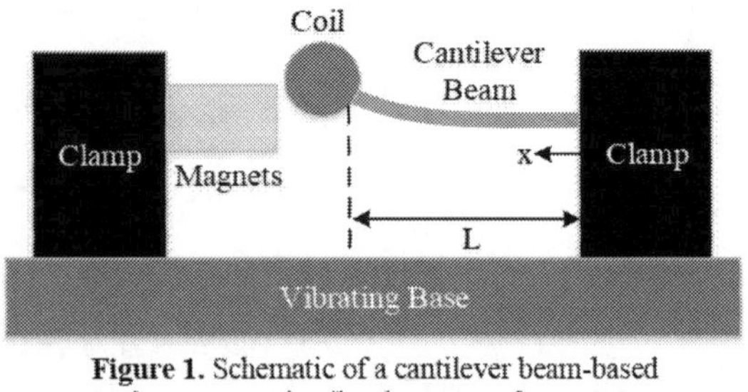

Figure 1. Schematic of a cantilever beam-based electromagnetic vibration energy harvester.

Based on Faraday's law of electromagnetic induction, power can be generated when the cantilever beam vibrates, causing the coil to cut through the magnetic field of the magnets. To fully model the effect of damping on the power output of the harvester, the governing equations relating the harvester's mechanical damping and electromagnetic damping to its power output must be considered simultaneously.

2.1 Mechanical damping

Assuming linear damping, the mechanical damping ratio of a vibrating structure can be related to its maximum stress by the following equation [6]

$$\zeta_m = 3619.5(\sigma_m^{0.3}/\sigma_f^{-2.3}) + 2887.1(\sigma_m^{6}/\sigma_f^{-8}) \quad (1)$$

where ζ_m is the mechanical damping ratio of the structure, σ_m is the maximum stress experienced by the vibrating structure and σ_f is the fatigue limit stress of the structure. Erturk and Inman [8] modelled the equation of motion of a clamp-free cantilever beam with proof mass under harmonic base excitation for first mode vibrations as the following

$$z(x) = K(x)Y_0/2\zeta_m \quad (2)$$

where $z(x)$ is the relative amplitude of the beam at position x along the length of the beam, $K(x)$ is a constant derived from the inertial and static moment terms of the beam and its proof mass, Y_0 is the harmonic base excitation amplitude and ζ_m is the mechanical damping ratio of the beam. Here, the proof mass is assumed to be rigidly fixed onto the free-end of the beam. By applying the Euler-Bernoulli's bending theory, the maximum stress of the vibrating beam can be determine by equating $x = 0$.

$$\sigma_m = EhK''(0)Y_0/4\zeta_m \quad (3)$$

where E is the Young's modulus of the beam, h is the beam's thickness and $K''(0)$ is the second derivative of $K(x)$ with respect to x, evaluated at $x = 0$. By substituting equation (3) into equation (1), the mechanical damping ratio of the beam can be determined analytically, provided that the material properties of the beam are known.

2.2 Electromagnetic damping

The electromagnetic damping of an electromagnetic vibration energy harvester is given by [9]

$$\zeta_e = (NBL_c)^2/2m_e w_n(R_c+R_L) \quad (4)$$

where N is the number of turns of coil, B is the strength of the magnetic field, L_c is the practical length of the coil, m_e is the effective mass of the harvester, w_n is the natural frequency of the harvester and R_c and R_L is the coil and load resistance. For a cantilever beam with a proof mass attached onto its free-end, the effective mass of the system can be approximated as [10]

$$m_e = m_p + 33/140 m_b \quad (5)$$

where m_p is the mass of the proof mass and m_b is the mass of the cantilever beam.

2.3 Power output

Kulah and Najafi [11] reported that the maximum power output of an electromagnetic vibration energy harvester at the load resistance can be expressed as the following

$$P_{max} = (NBL_c v)^2\, R_L \,/\, (R_c+R_L)^2 \quad (6)$$

where P_{max} is the maximum load power output of the harvester and v is the peak velocity of the coil cutting though the magnetic field. Considering both mechanical and electromagnetic damping and assuming that the peak velocity is equal to the maximum velocity at the free-end of the cantilever beam where $x = L$, v can be defined as

$$v = w_n\, K(L)Y_0/2(\zeta_T) \quad (7)$$

where ζ_T is known as the total damping and is equal to the sum of the mechanical and the electromagnetic damping. It is easy to notice that the velocity term in equation (7) is actually a product of equation (2) and the beam's natural frequency.

3 Results and discussion

Equation (4) suggest that increasing the effective mass of the harvester would decrease the overall electromagnetic damping. However, this will was also lead to an increase in stress and mechanical damping as described by equation (1). The effect of effective mass on the total damping of an electromagnetic vibration energy harvester was investigated in this section. Here, two cases were considered with the first case assuming a constant optimum load resistance and the second case considering the change in optimum load resistance for every variation in effective mass. These two cases were investigated due to the different theory on optimum load resistance by previous authors.

3.1 Case 1 – Constant optimum load resistance

Yang et al. [12] reported that the optimum load resistance of an electromagnetic vibration energy harvester is equal to the coil resistance and unaffected by other parameters. Therefore, increasing the effective mass of the harvester would not change the optimum load resistance value. Table 1 states the parameters used in the analysis.

Table 1. Parameters used in the theoretical analysis.

E (GPa)	69
N	400
L_c (mm)	4.0
R_c (Ω)	6.7
σ_f (MPa)	95.6

The material properties of an aluminium beam was considered in Table 1 as equation (1) is only valid for metals [6]. A constant base excitation acceleration of 0.1g was applied. Figure 2 describes the variation in electromagnetic and mechanical damping with effective mass for different magnetic field strengths, B. Figure 3 describes the corresponding total damping from Figure 2. Here, the effective mass term was translated into the mass ratio, m_r, of the proof mass to the beam mass as this will give a more general insight on the analysis.

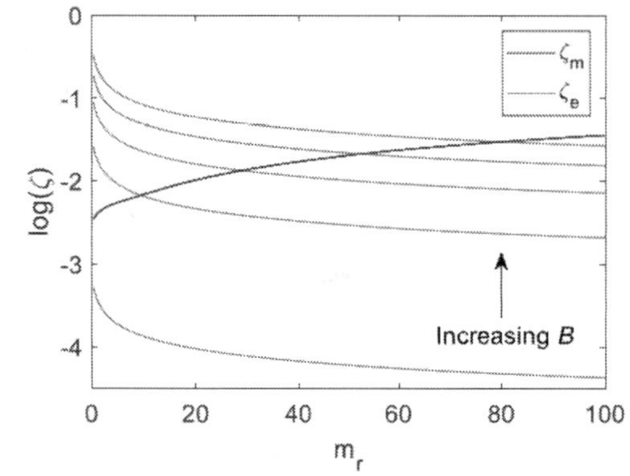

Figure 2. Case 1 variation in mechanical and electromagnetic damping ratio with mass ratio.

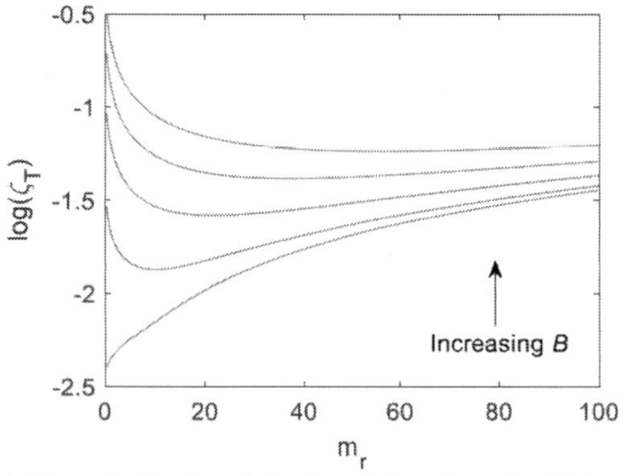

Figure 3. Case 1 variation in total damping ratio with mass ratio.

Figure 2 shows that increasing the mass ratio of the harvester results in an increase in mechanical damping and a decrease in electromagnetic damping. The results in Figure 3 illustrates that at regions where the electromagnetic damping is higher than the mechanical damping, the total damping decreases when the mass ratio increases. However, an opposite trend is seen when the electromagnetic damping is smaller than the mechanical damping. Overall, this analysis suggest that increasing the mass ratio of the electromagnetic harvester is favourable when the electromagnetic damping is larger than the mechanical damping, as this will decrease the total damping and hence an increase in gain. However, equation (6) suggest that a decrease in total damping does not necessarily lead to an increase in power as there are other parameters that must be considered such as the natural frequency on the harvester. Increasing the mass ratio would also result in a decrease in the natural frequency. Therefore, the output power was plotted against the variation in mass ratio to investigate this matter. Figure 4 shows the variation in the maximum load power for each different mass ratio input. Each curve in Figure 4 corresponds to the results in Figures 2 and 3.

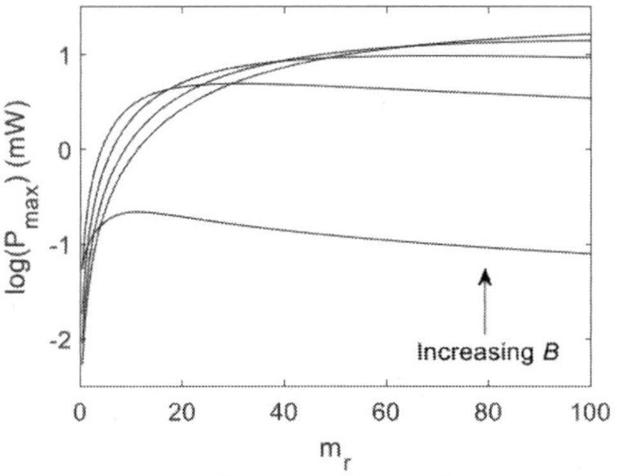

Figure 4. Case 1 variation in maximum load power output with mass ratio.

Figure 4 shows that an optimum mass ratio exist where the harvester would generate a maximum load power output. This optimum value increases when the magnetic field strength increases. This analysis also suggest that for every different harvester designs, the harvester has an optimum operational natural frequency. However, for cases of large optimum mass ratios, practicality may become an issue.

3.2 Case 2 – Varying optimum load resistance

Saha et al. [13] stated that when the electromagnetic damping of an electromagnetic vibration harvester becomes significant, the optimum load resistance does not equal to the coil resistance. The analysis in Case 1 was repeated by considering the change in optimum load resistance for every variation of effective mass.

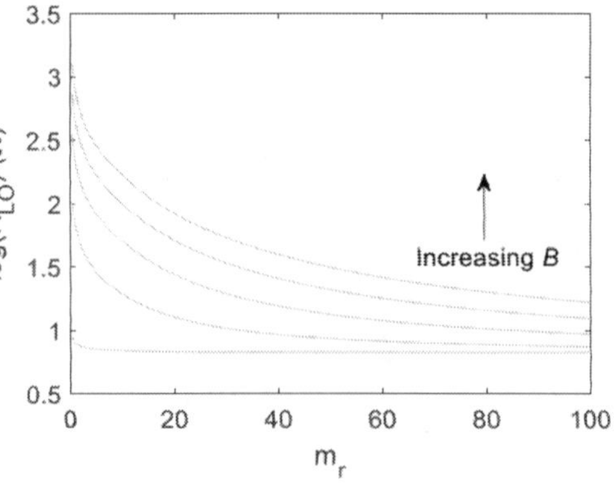

Figure 5. Variation in optimum load resistance with mass ratio

All analysis in Case 2 uses the same parameters applied as in Case 1. Figure 5 describes the variation in optimum load resistance and mass ratio. The optimum load resistance was determine using equation (6), in where the optimum value was selected based on the load resistance that resulted in the maximum load power output. It is shown that increasing the mass ratio of the harvester decrease the optimum load resistance value. Figure 6 shows the variation in mechanical damping and electromagnetic damping for Case 2 and Figure 7 describes the corresponding total damping.

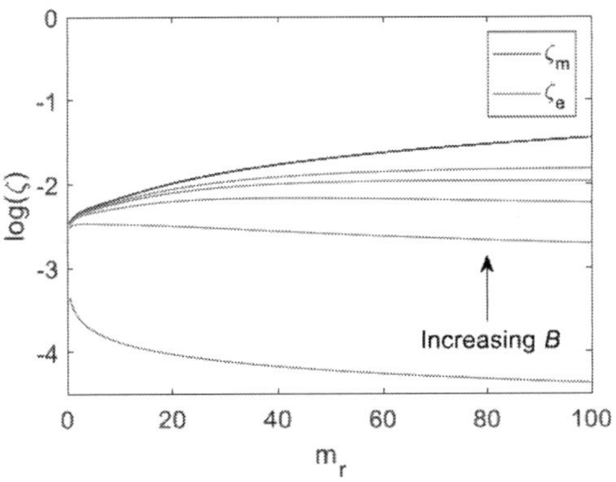

Figure 6. Case 2 variation in mechanical and electromagnetic damping ratio with mass ratio.

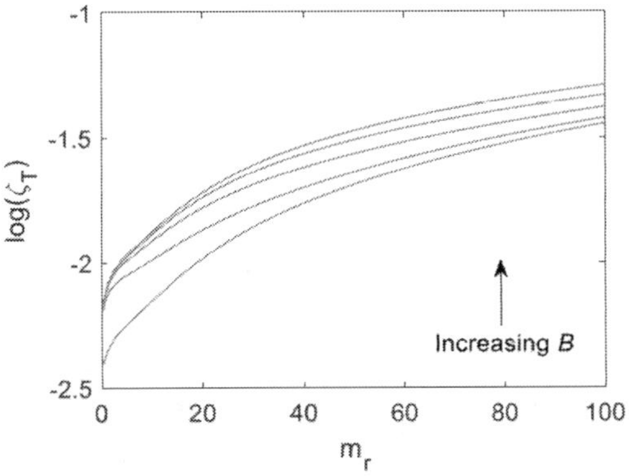

Figure 7. Case 2 variation in total damping ratio with mass ratio.

Analytical results in Figures 6 and 7 portrays a considerably different result than Case 1. It is shown that when the variation in optimum load resistance for every different mass ratio is considered, the electromagnetic damping will always be lower than the mechanical damping for any mass ratios, even under high magnetic field strength. This resulted in an increasing trend in the total damping for all magnetic strength variation as seen in Figure 7. Nevertheless, the power output results for Case 2 is somewhat similar to Case 1. Figure 8 shows that there also exist an optimum mass ratio or natural frequency in where the harvester would generate maximum load power. It worth to note that if the mechanical damping was assumed to be constant for all mass ratios, then the optimum mass ratio shown in Figures 4 and 8 would not exist, as the load power output would display a continuously increasing trend with when the mass ratio increases. However, this assumption is an over-generalisation for cantilever beam-based harvesters and may lead to large errors when performing large-scale optimisation.

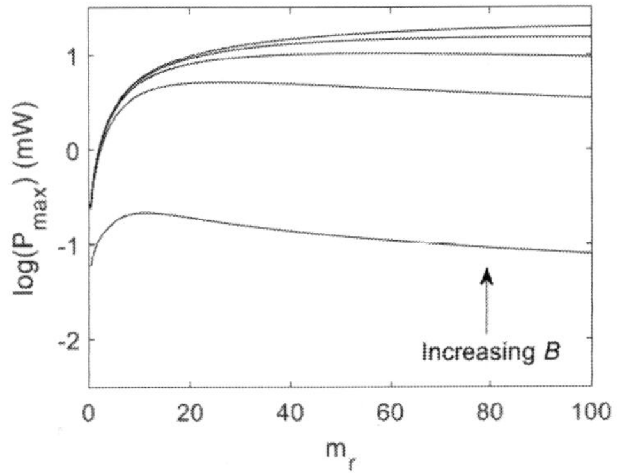

Figure 8. Case 2 variation in maximum load power output with mass ratio.

4 Conclusion

This study presents a theoretical analysis on the influence of total damping on the power output of an electromagnetic vibration energy harvester. Initially, the governing equations for the equation of motion and the power output of a cantilever beam-based electromagnetic vibration energy harvester were derived. Both variations in mechanical damping and electromagnetic damping for different harvester specifications were considered based on models developed from past literatures. Analytical results shown that increasing the effective mass of a cantilever beam-based harvester would result in an increase in the mechanical damping and a decrease in the electromagnetic damping. The total damping of the harvester displayed an increasing trend with effective mass at regions where the electromagnetic damping is lower than the mechanical damping. An opposite trend was recorded when the electromagnetic damping is larger. If the change in optimum load resistance was considered for every different effective mass variation, the electromagnetic damping will always fall below the mechanical damping. In terms of power output, results described that there is an optimum mass ratio value where the harvester would produce maximum power. This optimum value varies according to the harvester's design specifications. Hence, this also suggest that different harvester designs would have a different optimum operational natural frequency. Future works include deriving the exact expression for the condition of optimum mass ratio and verifying the results through experimental methods. The analytical analysis presented in this work are only valid for electromagnetic vibration energy harvesters, and exploration into damping analysis for other types of vibration energy harvesters may be considered in the future.

Acknowledgement

The authors would like to thank the Fundamental Research Grant Scheme (FRGS) from Ministry of Higher Education (MOHE) Malaysia, Grant No:

FRGS/1/2016/TK03/HWUM/03/1, for funding this research.

References

[1] S. Dua,, Y. Jia, S. Chena, C. Zhaoa, B. Sunc, E. Arroyoa, A.A. Seshiaa, Sens. Actuat. A, **263** (2017)

[2] C.K. Thein, J.S. Liu, Eng. Comput., **33** (2017)

[3] C. Wei, X. Jing, Renew. Sustain Energy Rev., **74** (2017)

[4] S.G. Kim, S. Priya, I. Kanno, MRS Bulletin, **37.11** (2012)

[5] Y. Tan, Y. Dong, X. Wang, J. Microelectromech., **26(1)** (2017)

[6] B.J. Lazan, *Damping of materials and members in structural mechanics* (Pergamon Press, London, 1968)

[7] E. Arroyo, A. Badel, Sens. Actuat. A, **171** (2011)

[8] A. Erturk, D.J. Inman, J. Intel. Mater. Syst. Struc., **19** (2008)

[9] B.L. Ooi, J.M. Gilbert, Sens. Actuat. A, **213** (2014)

[10] H. Yu, J. Zhou, L. Deng, Z. Wen, Sens., **14** (2014)

[11] H. Kulah, K. Najafi, IEEE Sens. J., **8(3)** (2008)

[12] B. Yang, C. Lee1, W. Xiang, J. Xie1, J. Han He, R.K. Kotlanka, S.P. Low, H. Feng, J. Micromech. Microeng., **19** (2009)

[13] C.T. Saha, T. O'Donell, H. Loder, S. Beeby and J. Tudor, IEEE Trans. Magnetics, **42(10)** (2006)

Transient Dynamic Analysis of a Cantilever Rod with Axial Impulse Loading Using Finite Element Method

Sushmita Deka , Pallekonda Ramesh Babu and Maneswar Rahang

National Institute of Technology Meghalaya, India

Abstract. The behavior of bodies subjected to impulse loading is of prime importance in the study of forces that occur in impulse facilities. Before performing the actual tests, theoretical and numerical simulations are carried out to obtain the response of bodies subjected to impulse loading. The simplest model for this study can be considered as a rod of circular cross section fixed at one end and free at other end. When a transient impulse load is applied on a body, vibrations occur in the body for the brief period of time. In this paper, the effect of a half sine impulse force applied on a cantilever rod in the axial direction has been discussed. The displacements at the tip of the rod were obtained based on two theories, the basic vibration formulae and FEM analysis. Simulations were performed using ANSYS and compared with the displacements obtained from the two theoretical methods.

1 Introduction

Impulse facilities like wind tunnels and shock tunnels are used for testing of various scaled aerodynamic models to find the forces and moments which it undergoes when subjected to various types of loads. These facilities are of short test duration in the order of milliseconds. The theoretical and numerical analysis to find forces and moments are generally carried out before actual testing in shock tunnels and wind tunnels. These analysis are usually performed by Computational Fluid Dynamics (CFD) and Finite Element (FE) simulations. However, the vibrations occurred in the body due to the applied forces can also be studied to know the effect of the forces on the body. The vibrations result in deformation of the body, and the forces can be predicted from various techniques by analyzing the deformation occurring on the body as a result of the applied force. Some of the researchers have worked on determining the response of the system subjected to various kinds of loads. Jia-shi *et al.* [1] studied the longitudinal forced vibration of a rod with non-linear boundary conditions having a concentrated mass at one end and supported by a spring on the other end. Wei [2] reviewed the vibration analysis of strings, rods, beams, membranes and thin plates using the discrete singular convolution algorithm. Temel and Calim [3] studied the forced vibration of cylindrical helical cantilever rod subjected to impulse loading in the Lapalce domain using the complementary functions method. The Timoshenko beam theory was used to derive the governing equations for cylindrical helical rods. The displacements and moments obtained were compared with those obtained from ANSYS simulations. Arani *et al.* [4] calculated the natural frequencies and mode shapes of rods with variable cross section considering various boundary conditions like clamped-clamped, clamped-free, free-free and free-clamped. They used a wave propagation method to find the natural frequency and mode shapes and compared it with the analytical method. A higher accuracy was obtained from the wave propagation method than the analytical method. Shokrollohi and Nejad [5] studied the longitudinal free vibration of non-uniform clamped-clamped and clamped-free rods to find the natural frequency by using discrete singular convolution. Kruti [6] studied the free longitudinal vibrations of a straight rod fixed at one end to find the natural frequencies of the rod. The rod was considered with uniform properties throughout the length of the rod and the method of force integration was used. Mei [7] studied the vibrations in uniform and stepped rods using four rod theories, the elementary theory, the Love theory, the Mindlin-Herrman theory and the three mode theory in which the motion of the rod is described by the propagation and reflection of waves through the rod. Shali *et al.* [8] studied the vibration of non-uniform rods having clamped-clamped and clamped-free ends to find the natural frequency using differential transform method by solving ordinary and partial differential equations. Collini *et al.* [9] studied the vibration analysis of tie rods having elastic bed type boundary conditions and compared with a model having both end fixed boundary conditions. The natural frequency of the tie rod was found from which the axial loads on the tie rod was determined. The vibrations on a rod due to an applied force were also studied using finite element analysis by researchers. Mottershead [10] studied the vibrations of a helical rod using finite element analysis to find the natural frequency of the helical rod. The frequency obtained was compared with those of a helical spring. Karami *et al.* [11] studied the free vibrations of spatial rods fixed at one end using finite element analysis. Adhikari *et al.* [12] studied the free and forced axial vibrations of rods with non-local elasticity using finite element analysis.

Although a considerable amount of work has been carried out on the study of vibration of cantilever rods, very limited work has been reported in literature in comparing the response using analytical techniques such as vibration formulae and the finite element method. In this paper, a solid rod of circular cross section made of aluminium has been considered for the analysis. Axial half sine impulse force has been considered to apply on the model to find the displacement at the tip of the rod due to the force. The analysis is performed using vibration formulae and FEM analysis along with simulations in ANSYS. The displacements obtained from the three techniques were compared with each other. The schematic diagram of the cantilever rod and a half sine impulse force is shown in Figure 1(a) and Figure 1(b).

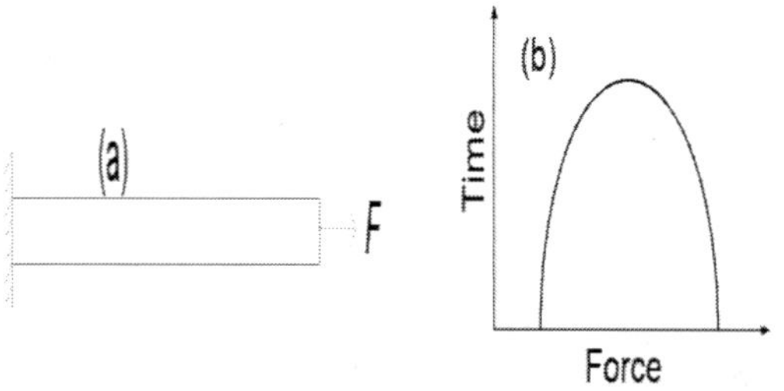

Figure 1: (a) Schematic diagram of a cantilever rod (b) Half sine impulse force

2 Formulation

To find the displacement at the tip of the cantilever rod subjected to impulse load, the vibration theory and the FEM theory has been used. The displacement at the tip of the cantilever rod is measured for transient loading. The transient load is considered to follow a harmonic behavior with very short time duration in the order of milliseconds. The force is acting at the end of the rod in the axial direction i.e. no deformation occurs in the y and z direction. The rod cross section is considered to be circular and the displacement at the tip of the rod is found for various lengths and loads on the rod. The displacement at the tip of the rod is also found for different values of time for transient loading. Let the length of a cantilever rod be L on which an axial transient load $f = F \sin \omega t$ is acting at the end of the rod, where F is the amplitude of the load, ω is the frequency of the load and t is the time for which the load is acting. Let the stiffness of the rod be k and the displacement at the tip of the rod be denoted by u. For a constant force F,

Force (F) = Stiffness (k) × displacement (u)

i.e. $$u = F/k \tag{1}$$

Here E is the Young's modulus, A is the cross section area of the rod and l is the length of each element of the rod. The generalized stiffness and force matrices considering 3 elements of the rod are represented as given below using FEM,

$$k = \frac{EA}{l}\begin{bmatrix} 1 & -1 & 0 & 0 \\ -1 & 2 & -1 & 0 \\ 0 & -1 & 2 & -1 \\ 0 & 0 & -1 & 1 \end{bmatrix} \tag{2}$$

$$F = \begin{bmatrix} F_1 \\ F_2 \\ F_3 \\ F_4 \end{bmatrix} = \begin{bmatrix} 0 \\ 0 \\ 0 \\ F \end{bmatrix} \tag{3}$$

Displacement at each node,

$$\begin{bmatrix} u_1 \\ u_2 \\ u_3 \\ u_4 \end{bmatrix} = \frac{EA}{l}\begin{bmatrix} 1 & -1 & 0 & 0 \\ -1 & 2 & -1 & 0 \\ 0 & -1 & 2 & -1 \\ 0 & 0 & -1 & 1 \end{bmatrix}^{-1} \begin{bmatrix} 0 \\ 0 \\ 0 \\ F \end{bmatrix} \tag{4}$$

The force considered in the present condition is transient impulse force in the form, $f = F \sin \omega t$.

The displacements with respect to time at any node denoted as U and the expression for the same is given by

$$U = k^{-1} F \sin \omega t \tag{5}$$

where t varies from zero to the total time.

According to vibration theory, the equation of motion of a rod in axial direction is given by

$$\frac{\partial}{\partial x}\left(EA\frac{\partial u}{\partial x}\right) = \rho A \frac{\partial^2 u}{\partial t^2} \tag{6}$$

where ρ is the density of the material of the rod.

If the property of the rod do not change with length, then

$$\frac{\partial^2 u}{\partial t^2} = c^2 \frac{\partial^2 u}{\partial x^2} \tag{7}$$

where $c^2 = \frac{E}{\rho}$ is the speed of sound in the rod. The equation (7) is the one dimensional wave equation which describes the propagation of electromagnetic waves in a homogenous medium. The boundary conditions of a cantilever rod fixed at one end are

$$u(0,t) = 0\ ,\ \frac{\partial u}{\partial x}(l,t) = 0$$

Equation (7) can be solved as an eigen value problem for vibration of a rod using method of separation of variables i.e.

$$u(x,t) = U(x)T(t) \tag{8}$$

which gives the following equation

$$\frac{1}{T(t)}\frac{d^2T}{dt^2} = c^2\frac{1}{U(x)}\frac{d^2U}{dx^2} \tag{9}$$

Let each of the terms in equation (9) be equal to a constant $-\omega^2$. Thus

$$\frac{1}{T(t)}\frac{d^2T}{dt^2} = -\omega^2 \tag{10}$$

And $$c^2\frac{1}{U(x)}\frac{d^2U}{dx^2} = -\omega^2 \tag{11}$$

On solving the equations (10) and (11) and applying the boundary conditions, the natural frequency (ω_n) of the cantilever rod is given the following formula [13],

$$\omega_n = \frac{\pi\sqrt{E/\rho}}{2L} \tag{12}$$

For transient impulse force, $f = F \sin \omega t$, displacements with respect to time during the pulse duration at any node is given by [14],

$$U = \frac{F/k}{1-\left(\frac{\omega}{\omega_n}\right)^2}\left[\sin(\omega t) - \left(\frac{\omega}{\omega_n}\right)\sin(\omega_n t)\right] \tag{13}$$

where t varies from zero to the total time.

After the pulse, the displacement with respect to time is given by [14],

$$U = \frac{F/k}{1-\left(\frac{\omega}{\omega_n}\right)^2}\left(\frac{\omega}{\omega_n}\right)\left[\sin(\omega_n t) - \sin(\omega_n (t-t'))\right] \tag{14}$$

where t' is the impulse duration.

3 Results and discussion

3.1 Effect of the length of the rod on the displacement at the tip

The half sine impulse force $f = F \sin \omega t$, having an amplitude of 10 N has been applied on the aluminum rod with 0.01 m diameter for a time of 1 ms. The displacement at the tip of the cantilever rod is plotted for three different lengths i.e. 0.2 m, 0.5 m and 0.8 m. The displacement history obtained using the analytical FEM and vibration theory for the applied forces are shown in Figure 2(a) and Figure 2(b).

The displacement at the free end of the rod increases with increase in the length. Since the stiffness of a body is inversely proportional to the length, hence the displacement at the end of the rod increases with increasing length.

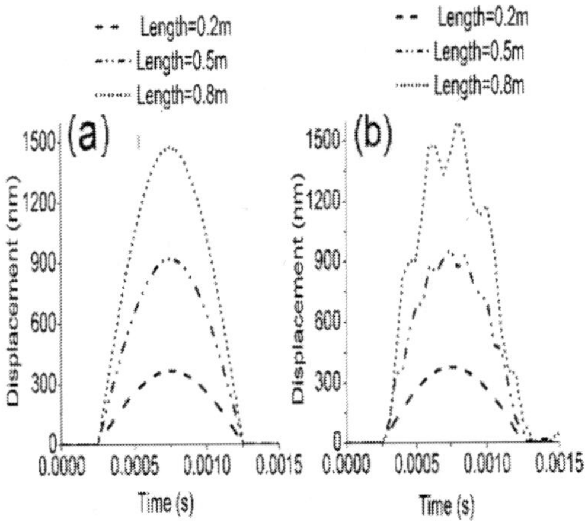

Figure 2: (a) Displacement history obtained from analytical FEM method of the rods with different lengths. (b): Displacement history obtained from vibration analysis of the rods with different lengths.

Numerical simulations were performed in ANSYS using finite element method for different lengths of the rod and is compared with the displacements obtained from analytical FEM and vibration analysis. The parameters were kept the same as those used for the two techniques. One end of the rod was fixed while the other end was free and the displacement was considered only in the axial direction. Hexahedral elements were used for the simulation and the number of element were 560. The displacements obtained from the ANSYS simulations for three given lengths are given in Figure 3. The obtained displacements agree well with the displacements obtained from the theoretical techniques. The displacements obtained from FEM, vibration and ANSYS simulations for 0.2 m length of rod are compared in Figure 4.

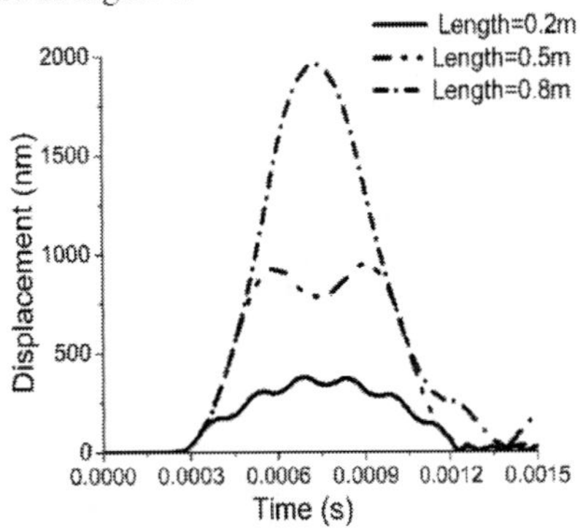

Figure 3: Variation of displacement with time at three different lengths obtained using ANSYS

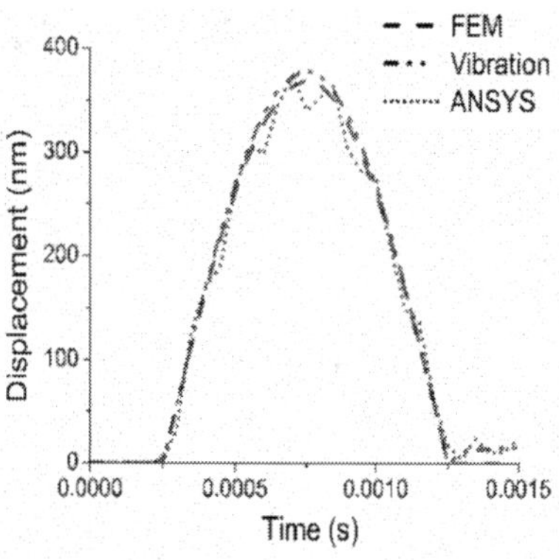

Figure 4: Variation of displacement with time for 0.2 m length of rod obtained using FEM, vibration and ANSYS

3.2 Displacement at the tip of the cantilever rod on varying the load

The displacement history obtained at the tip of the cantilever rod was plotted for three different amplitude of loads applied at free end of the rod, i.e. 10 N, 20 N and 30 N. The force is considered to act harmonically following the equation $f = F \sin \omega t$ for a time of 1 ms. The other parameters were kept the same for three cases. The graphs obtained using the FEM analysis and vibration theory are shown in Figure 5(a) and Figure 5(b).

The displacement at the free end of the rod increases with increase in amplitude of load. The displacements obtained from the FEM analysis and vibration analysis are found to agree well with each other.

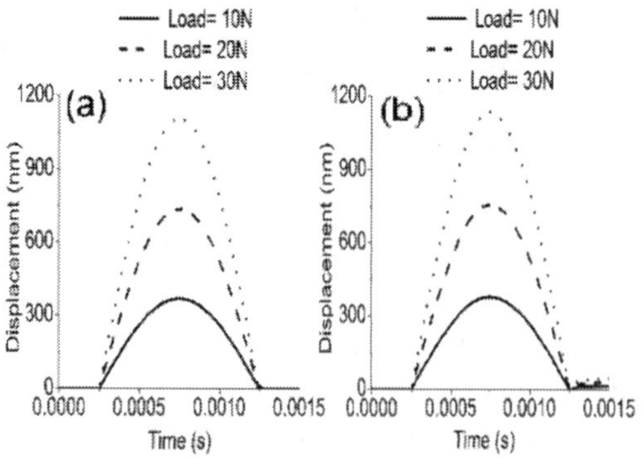

Figure 5 : (a)Variation of displacement with time at three different amplitude of loads obtained from FEM analysis

(b):Variation of displacement with time at three different amplitude of loads obtained from vibration analysis

Using the same ANSYS model as used earlier, the simulations were performed to find the displacement at the free end of the rod for three different amplitude of loads and the displacements were found to agree well with the theoretical results. The displacements obtained from the simulation are shown in Figure 6.

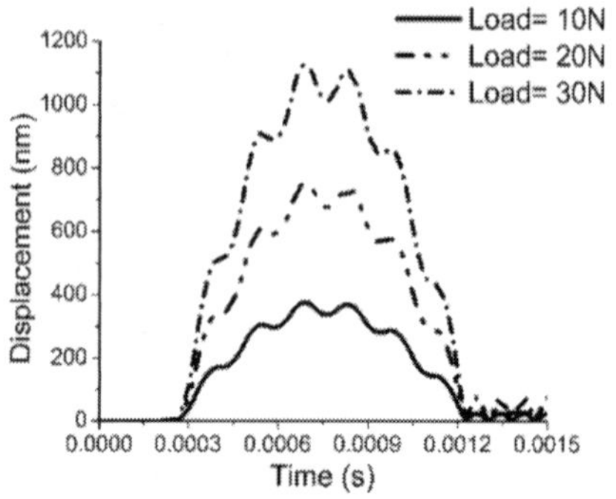

Figure 6: Variation of displacement with time at three different loads obtained using ANSYS

3.3 Displacement at the tip of the cantilever rod on varying the total time of impulse

The total time for which the harmonic load is applied is varied for three different times- 1 ms, 2 ms and 3 ms. The effect of the varying impulse time on the displacement at the free end of the cantilever rod is analyzed keeping other parameters like length of the rod and amplitude of the load as constant for the three cases. The length of the rod is considered as 0.2 m and amplitude of the load applied is considered as 10 N. The graphs obtained using the FEM analysis and vibration theory are shown in Figure 7. It is found that the maximum displacement at the tip of the rod remains same for all the three duration of impulses for both the theoretical methods. Both the theoretical methods to find the displacement are found to agree well with each other.

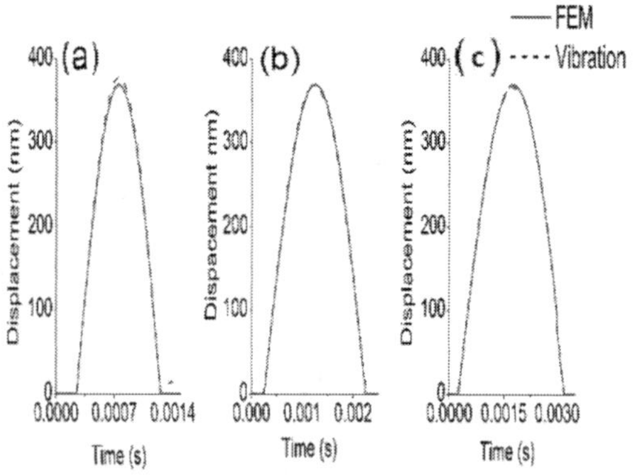

Figure 7: Variation of displacement with time at three different impulse durations obtained from FEM analysis and vibration analysis

The same ANSYS model is again used to perform simulations to analyze the effect of different impulse duration on the displacement at the tip of the cantilever rod. The displacements obtained agreed well with those obtained from theoretical analysis. The maximum displacement is almost same for all the three time durations. The displacements obtained from the ANSYS simulations are shown in Figure 8.

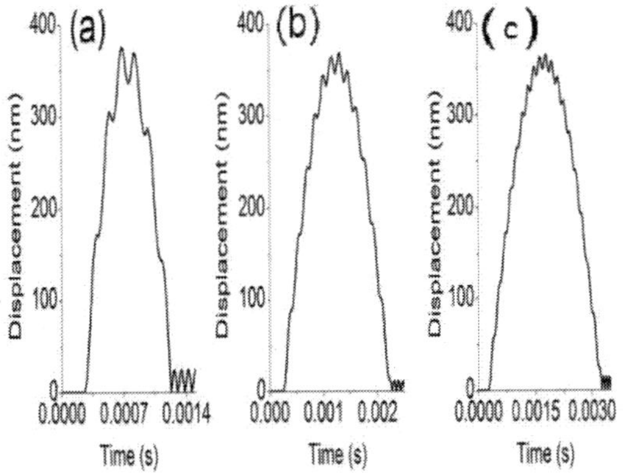

Figure 8: Variation of displacement with time at three different impulse durations obtained from ANSYS

4 Conclusion

In the present study, a half sine impulse force is applied on a cantilever rod made of aluminium and the displacement is measured at the tip of the rod for different lengths, different amplitudes of loads and different impulse duration. The analytical FEM theory and the vibration theory was used to measure the displacement at the tip of the rod. The displacements obtained from both the methods are compared with each other and good agreement has been found between the two theories. The displacements are also compared with the displacements obtained from ANSYS simulations and they were found to agree well with those obtained from theoretical analysis. However, the displacements obtained from FEM and vibration theory were found to be smoother compared to that obtained from the ANSYS simulations.

References

1. T. Jia-shi, L. Li and H. Quan-zhong, Vibration analysis of a rod with complex boundary conditions, Appl. Math. Mech. **9**(1988), 837-847
2. G. W. Wei, Vibration analysis by discrete singular convolution, J sound Vib. **244**(2001), 535-553
3. B. Temel and F. F. Calim, Forced vibration of cylindrical helical rods subjected to impulsive loads. J. Appl. Mech. **70**(2003), 281-291.
4. M. K. Arani, N. R. Saleh and M. N. Bahrami, A Modified Wave Approach for the Calculation of Natural Frequencies and Mode Shapes of the Rods with Variable Cross Section, App. Mech. Mat. **110**(2012), 2537-2547
5. M. Shokrollahi, and A. Z. B. Nejad, Numerical analysis of free longitudinal vibration of non-uniform rods: Discrete singular convolution approach, J ENG MECH ASCE. **140** (2014), 06014007
6. Y. Krutii, Analysis of longitudinal oscillations for systems with continuous variable parameters using force integration method, Tehnicki glasnik, **9**(2015), 420-425
7. C. Mei, Comparison of the four rod theories of longitudinally vibrating rods, J. Vib. Control. **21**(2015), 1639-1656
8. S. Shali, S. R. Nagaraja, and P. Jafarali, Vibration of non-uniform rod using Differential Transform Method, Mater. Sci. Eng. **225**(2017), 012027
9. L. Collini, R. Garziera, and K. Riabova, Vibration Analysis for Monitoring of Ancient Tie-Rods, SHOCK VIB. 2017
10. J. E. Mottershead, Finite elements for dynamical analysis of helical rods, Int. J. Mech. Sc. **22**(1980), 267-283.
11. G. Karami, M. Farshad, and M. Yazdchi, Free vibrations of spatial rods—A finite-element analysis, Int J Numer Method Biomed Eng. **6**(1990), 417-428.
12. S. Adhikari, T. Murmu, and M. A. McCarthy, Dynamic finite element analysis of axially vibrating nonlocal rods, FINITE ELEM ANAL DES. **63**(2013), 42-50.
13. W.T. Thomsn, M.D. Dahleh, C. Padmanabhan, *Theory of Vibration with Applications*. 5thedn, Pearson Education (1998)
14. L. Meironvtch, *Fundamentals of Vibration*, Waveland Press (2010)

Chatter Behavior in the Milling Process of Inconel 718: Effects of Tool Edge Radius

Hoe C H, Reddy M M , Lee V C C and Debnath S

Curtin University Malaysia, CDT 250, 98009 Miri, Sarawak, Malaysia

mohan.m@curtin.edu.my

Abstract. Inconel 718 is widely used in various high end industries such as aerospace, nuclear plant, petrochemical plants etc. Inconel 718 is used for these applications due to unique mechanical properties such as high mechanical strength at elevated temperatures, high resistance to corrosion, and high strength to weight ratio. The unique properties of Inconel 718 made it difficult to be machined due to rapid work hardening and high cutting temperature. In addition, chatter vibration further increases the difficulty in machining of Inconel 718. In this paper, an experimental study on the effects of tool edge radius to the chatter behaviour was investigated. The dynamic responses of the milling process were recorded and analysed in both time domain and frequency domain. The results showed the variable helix and pitch end mill tool with larger tool edge radius able to mitigate chatter vibration at lower cutting speeds. Variable helix and pitch end mill with specific tool edge radius able to mitigate chatter vibration under the same cutting parameters. Experiments shows proper selection of tool edge radius improves the stability of end milling machining process.

1 Introduction

Inconel 718 is one of the Nickel based super alloys that are widely used in the various industries such as aerospace, nuclear, petrochemical plants, food processing equipment and etc [1-3]. Inconel 718 is famous in high temperature application owning to its outstanding properties such as able to maintain high mechanical strength at elevated temperature, high resistance to chemical wear, high resistance to corrosion as well as high strength to weight ratio [4]. However, Inconel 718 is extremely difficult to machine. In fact, Inconel 718 as one of the lowest machinability rating compared to other alloys [5]. Moreover, the combination of high-temperature mechanical properties, poor thermal properties (low thermal conductivity and thermal diffusivity) and rapid work hardening are the main factors that classified Inconel 718 as difficult-to-cut material [6]. As a result, high cutting temperature is generated at the cutting zone during the machining process of Inconel 718. Many researchers have carried out various studies to enhance the machinability of Inconel 718 by adapting different kind of cutting tools (coating, geometry and tool material) with different combination of cutting parameters and different cutting conditions [6-9]. For example, Jawaid [6] study the cutting performance and wear characteristic of coated and uncoated cutting tool in the face milling process of Inconel 718. The research shows that the coated tool has a better performance compared to the uncoated tool. Zhang [7] estimate the tool life and measure the cutting force in the end milling process of Inconel 718 under dry and minimum quantity cooling bio-lubrication condition. It is found that the tool life increases and has lower cutting force when the milling process is carry out under minimum quantity lubrication process. Hadi [8] compared the up and down milling process of Inconel 718 under different cutting condition, author found that the propagation of tool wear in up milling is more significant than down milling process. Although much effort has been work on to improve the machinability of Inconel 718, however, the chatter vibrations that develop during the machining process of Inconel 718 are rarely discussed in most of the studies regarding the machining process of Inconel 718.

Chatter vibration that caused by self-excited vibration is the most undesirable and the least controllable vibration. Taylor [10] described the chatter vibration is the "most obscure and delicate of all problems facing machinist". The self-excited vibration that occur in metal cutting are mainly due to the regenerative effect, the waviness of the surface that caused by the overlapping cutting process will further amplify the vibration and therefore, chatter vibration occurs [11,12]. Quintana [13] describe the adverse effect inflicted by chatter vibration

 https://doi.org/10.1051/matecconf/201820202006

can be devastating and therefore resulting in poor surface quality, machine tool damage, pre-mature tool wear, waste of energy, waste of material, and increase in production cost in term of production time, rapid tool wear. In order to avoid chatter vibration in the machining process, one of the methods is to generate the stability lobe diagram. The first stability lobe diagram was proposed by Tobias and Fiswick [14] for the turning process, stability lobe diagram is used to define the stable machining region (i.e no chatter) and the unstable machining region (i.e with chatter) as a function of axial depth of cut and spindle speed. The typical stability lobe diagram can be visualized in Figure 1 which formed by a series of intersected scallop-shaped lobe, the stable machining region is the area below the lobe while the unstable machining region is the area above the lobe. By selecting the machining parameters under the stability lobe can greatly improve the productivity of the machining process. In addition, the area on the left hand side of the stability lobe diagram is defined as the process damping region. The process damping only occur at low spindle speed, the damping phenomena occur at the contact between tool flank and the machined surface that will result in mitigation of chatter vibration [15]. However, the calculation of stability lobe diagram are difficult and lengthy process due to the modeling process of the multi-degree-freedom structures, there is a lot of parameters need to be considered in the development of stability lobe diagram such as the frequency response function of machine-tool, geometry of cutting tool, work piece material properties, radial immersion rate and so on. In addition, every stability lobe diagram is only unique for a specific process setup, the stability lobe need to be developed again if any of the parameters changes.

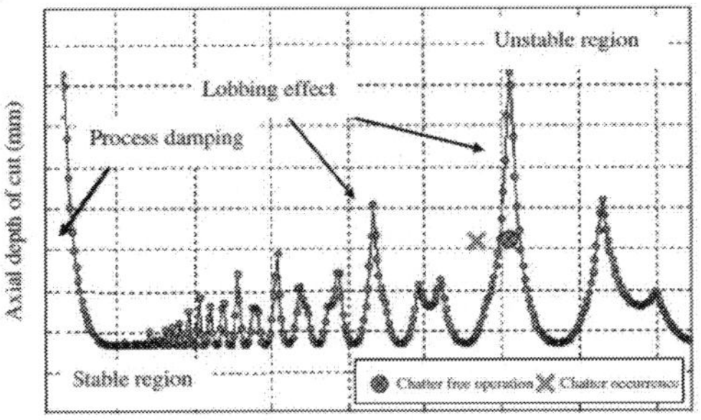

Figure 1 Stability lobe diagram [13]

Besides adapting stability lobe to avoid chatter vibration, another way to mitigate chatter vibration will be the use of variable helix and pitch end mill cutters which is first proposed by Slavicek [16]. The variable helix and pitch feature are designed to address the regenerative effect in the milling process by shifting the buildup of the chatter vibration. In other word, the regenerative effect is interrupted by the geometry of the cutting tool (ie. variable helix and pitch). Shirase [17] further shows that the variable helix and pitch end mill cutters able to reduce dimensional form errors as well as mitigate chatter vibration via experiment and simulation method. After that, Altintas [18] present an analytical model in predicting the stability lobes for variable pitch end mill cutters and the study demonstrated that variable pitch able to eliminate chatter vibration. In short, Literatures show that the variable pitch and helix end mill are capable to interrupt the dynamic system in the milling process and therefore increase the stability of the process [19-21]. Huang [19] observed lower vibration generated by variable pitch tool in the milling process of titanium alloy. Currently variable pitch and helix end mill cutters are available in the industrial environment. However, the use of variable pitch and helix end mill cutters is still not very extensive; this might due to the lack of knowledge about the cutters ability as well as its reliability. In addition, the variable helix and pitch end mills in the market now also available with multiple selection of tool edge radius. Study on the performance of the variable helix and pitch end mill cutters with tool edge radius end mill is limited.

The geometry of cutting tool edge play an important role in the machining process because the mechanical and thermal loads are directly sustain by cutting tool edge during the machining process [22]. Moreover, the tool life and the machined surface quality are affected by the shape of the cutting tool edge. There are three kind of cutting tool edge in general, namely (sharp, round and chamfer) and it is normally found in cutting tool with inserts. Endres [23] studied the effect of tool edge on the cutting tool flank wear in the turning process. The study concluded that there is a clear reduction of tool wear with the use of radius edge compare to the sharp edge as shown in Figure 2.

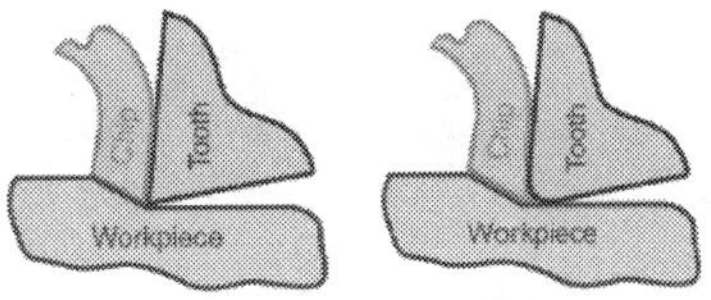

(a) (b)

Figure 2 Illustration of tool edge (a) Sharp tool (b) Radius edge

Therefore, one might hypothesize that the tool edge geometry of the cutting tool will also affect the chatter vibration in the milling process. In addition, the thermal load in the cutting tool edge should be able to reduce by spreading across the tool edge radius, potentially reduce the cutting temperature. This is especially important when machining a high heat resistance material such as Inconel 718. From the above discussion, variable pitch and helix cutting tools are able to mitigate the chatter vibration, and with the combination of tool edge radius, the cutting temperature reduces compare to the sharp cutting edge. The research articles related to the experimental investigations with variable pitch and variable helix angle milling cutters with tool edge radius is very limited in the literature especially in the end milling process of Inconel 718. In this paper, the variable pitch and helix end mill cutters are employed to study the effect of tool edge radius to the chatter vibration in the milling process of Inconel 718. Dynamics responses were

 https://doi.org/10.1051/matecconf/201820202006

recorded and analyzed during the milling process in the time domain and frequency domain to evaluate the chatter characteristic.

2 Experimental methods and materials

2.1 Experiment material and cutting tools

The work material used for the current experimental study is Inconel 718 with the dimension of 240 ×104 ×18 mm (l×b×h). The composition of the work piece is listed in Table 1.

Table 1 Inconel 718 composition

Material	Chemical composition (W_t) %									
	Ni	Cr	Nb	Mo	Ti	C	Si	Mn	B	Fe
Inconel 718	51.75	17	5.15	2.93	1.07	0.042	0.21	0.03	0.006	Last

The cutting tool used in the current experiment is a four flute solid carbide end mill with sub-micrograin size of 0.2μm and 12% cobalt contain which has high resistance to heat. The diameter of the cutting tool, (d) is 10mm and the total length of the cutting tool, (L) of 72mm as shown in Figure 3. The cutting tool is also chatter resistance with its unique geometry of variable helix angle (35° and 38°) and variable pitch angle (87°, 89°, 91°, 93°). In addition, the cutting tool also has tool edge radius as shown in Figure 3 which is denoted as R. Two cutting tools, namely cutting tool No.1 (0.3mm tool edge radius) and cutting tool No.2 (1.0mm tool edge radius) were used to carry out the experiment.

Figure 3 End mill cutters

2.2 Experiment setup and measurement

The experiment was performed on Leadwell V-30 vertical CNC machining center. The experiment setup can be illustrated in Figure 4. A tri-axial printed circuit board accelerometer is employed to capture the acceleration of the workpiece during the milling process. The accelerometer is adhesively mounted on top of the workpiece to sample the chatter vibration during the milling process. In order to keep the measurement consistent, the chatter vibration was sampled at the same location in every cutting pass. The chatter vibration was sampled at 2000Hz. The accelerometer is connected to data acquisition board (DAQ) and to the computer for data recording.

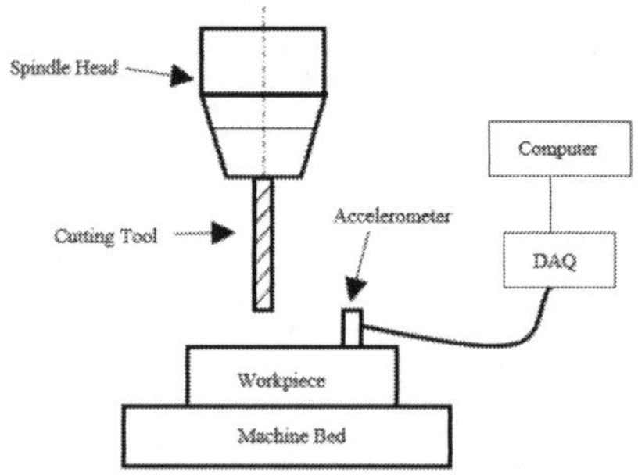

Figure 4 Experimental setup

2.3 Experiment parameters

The end milling process was carried out in dry condition with down milling mode. The cutting parameters for the experiments are listed in Table 2. Both cutting tool No.1 and cutting tool No.2 were utilized to perform milling process with different cutting speed.

Table 2

Parameters	Value
Cutting speed (m/min)	31.42, 62.83, 94.25, 125.67
Spindle Speed (RPM)	1000, 2000, 3000, 4000
Tool diameter (mm)	10
Axial depth of cut (mm)	0.1
Radial depth of cut (mm)	3
Table Feed (mm/min)	100
Tool overhang length (mm)	30

2.4 Modal analysis

The frequency response function (FRF) of the machine-tool setup was determined before the experiment, the purpose of finding the FRF of the machine-tool system is to ensure that the operating frequency will not trigger the resonance point of the system. The standardize procedure in obtaining the FRF of the system was carried out [25,26] . The impact test was carried out using a PCB-352C23 medium size steel-tip impact hammer and also PCB-352C03 ceramic shear ICP accelerometer. Both of the sensor data was collected by NI-USB 9234 4-channel data acquisition module. The result of the FRF can be illustrated at Figure 5, the first and second mode of the system are 860 Hz and 1280 Hz respectively. It is also note that the maximum operating frequency for the current experiment is around 267 Hz. Therefore, it should not trigger the resonance point of the machine-tool system. On the other hand, the workpiece-vice system is stiffer than the machine-tool system. Therefore, the FRF of the workpiece-vice system is not necessary.

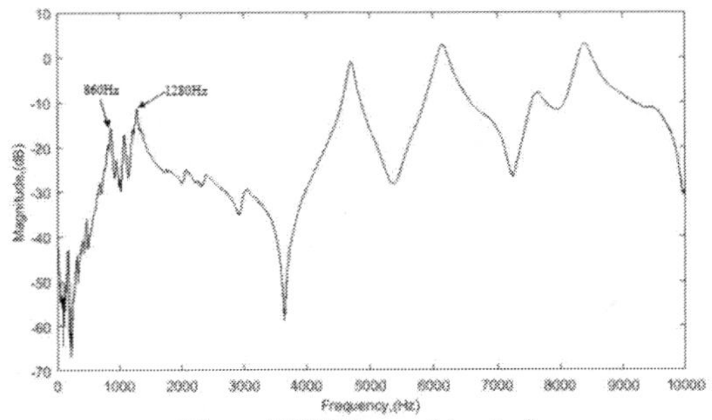

Figure 5 FRF for machine-tool

3 Results and discussion

3.1 Time domain analysis

In the milling process, the vibration in the x-axis, also known as cutting direction (illustrated at Figure 6) has dominant effect on the chatter vibration. Since the process is a down milling mode, the chip thickness is maximum upon entry and minimum at the exit. The high impact during the entry of the cutting zone is the starting point of the regenerative effect. This is the direction where the cutting tool edge exerts highest cutting force during the milling process. Figure 7 shows the typical time-domain response in the milling process of Inconel 718 at spindle speed of 1000 RPM. Based on the sampling rate and spindle speed, it can be calculated that each cutting cycle take about 120 data point at 1000 RPM. A cutting cycle is defined as one complete revolution of a cutting tool where all of the cutting edge are engaged with the work piece. The waveform shown in Figure 7 is at 1000 RPM, the waveform exhibits time periodic time-varying characteristics due to the intermittent cutting characteristic of the milling process. Also cyclic behavior of the chatter responses is observed, difference between the dynamic response for cutting tool No.1 and cutting No.2 is not similar. Cutting tool No. 1 shows a higher tendecy of vibration as compared to Cutting tool No. 2, where the response range of cutting tool No. 1 (p-p 0.4 g) is doubled of cutting tool No. 2 (p-p 0.2 g). It is also demonstrated that the responses for Cutting tool No. 2 attributes less noise, indicating a cleaner cut of work piece.

Further to the dynamic responses, a root mean square (RMS) value of the dynamic response is computed to evaluate the intensity of the chatter vibration.

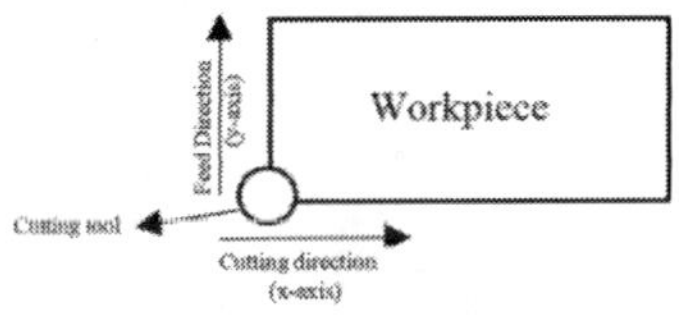

Figure 6 Cutting direction illustration

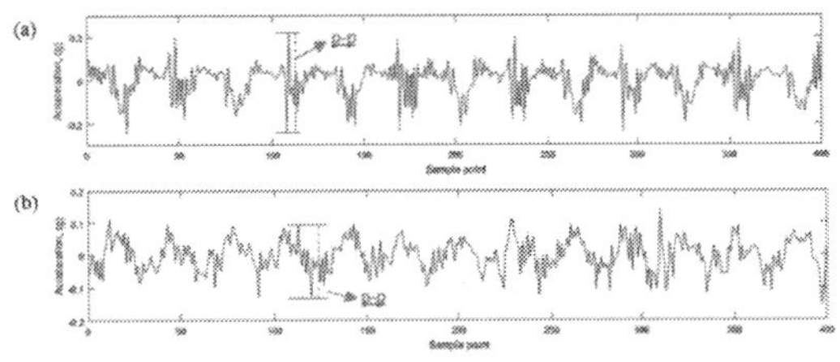

Figure 7 Time domain vibration response at 1000 RPM (a) Cutting tool No.1 (b) Cutting tool No.2

Figure 8 shows the RMS value of cutting tool No.1 and No.2 at different spindle speeds. The results show that the RMS value for cutting tool No.1 is higher than cutting tool No.2. The lowest RMS value are located at the lowest spindle speed for both cutting tools, this is due to the damping process effect. The stability of the milling process increases at lower cutting speed, in this case, the cutting speed is around 31m/s. Damping process phenomena has been explained by many of the researchers [14,15]. In addition to the process damping effect, the regenerative effect is also suppress by the unique geometry of the cutting tool (variable helix and pitch). Theoretically, the vibration could be higher with the standard end milling tool (uniform helix and pitch). In addition, it is also note that cutting tool No.1 (tool edge radius: 0.3mm) generally has higher RMS value compared to cutting tool No.2 (tool edge radius: 1mm). This is due to the ploughing force that is introduced by Albrecht [26], ploughing force is induced by the tool edge radius, and it can be considered as a frictional force that occur between the area at the flank face of the tool and the machined surface. The ploughing force increases with the increase of tool edge radius that is shown in the research work by Wyen [27], with the increase of ploughing force, it shows that the chatter vibration can be mitigated. Also the RMS acceleration amplitude is generally lower for cutting tool No.2 at most of the cutting speed except 3000 RPM. In other words, the initial guess is that the cutting tool with sharper cutting edge generate more noise, a representation of excessive vibrations and have a higher tendency to vibrate due to smaller of contact area between the flank wear and machined surface.

However, there is an exceptional at 3000 RPM, where cutting tool No.2 has higher value of RMS compare to cutting tool No.1 by around 8.8%. One possible reason is the cutting process has triggers chatter frequency of the cutting tool, hence creating high vibration. Considering 3000 RPM as the resonant frequency, at 4000 RPM, the cutting frequency would have surpass its resonance hence the tendency of high vibration is reduced. On the other hand, the process damping effect also diminishes with the increase of cutting speed, and therefore higher amplitude of vibration also expected. Another possible reason that causes the sudden increase in RMS value at 3000 RPM for both of the cutting tool may be due to the presence of work-hardening phenomena during the milling process. Work-hardening increases the resistance to the cutting process and induced higher intensity of vibration. Therefore, cutting tool No.1 also experience a sudden increase in RMS value from cutting speed of 2000 RPM to 3000RPM. The presence of work-hardening phenomena is due to the high cutting temperature as shown in Figure 9. In addition, the cutting temperature at 4000 RPM is too high that the cutting tool No.2 merely cutting the work piece with ploughing force just like grinding through the workpiece. Since cutting tool No.2 has higher value of tool edge radius, the contact area of the flank face and the machined surface is higher, and therefore the cutting edge is worn off due to high temperature cutting at 4000RPM as shown in Figure 10. This phenomena will be discussed when the signal is analysed in frequency-domain in the next section.

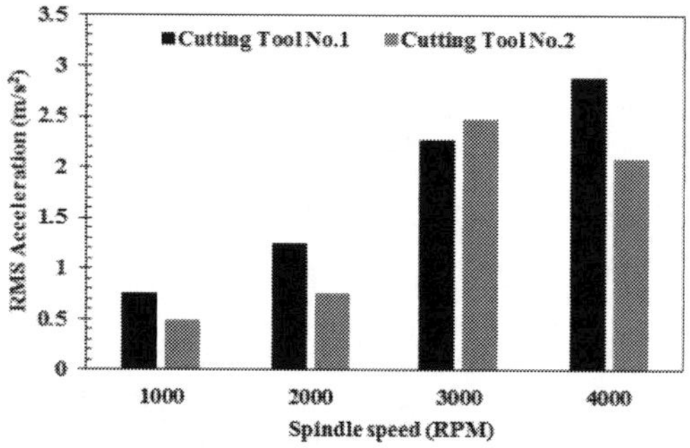

Figure 8 RMS acceleration value at various spindle speed

Figure 9 High cutting temperature

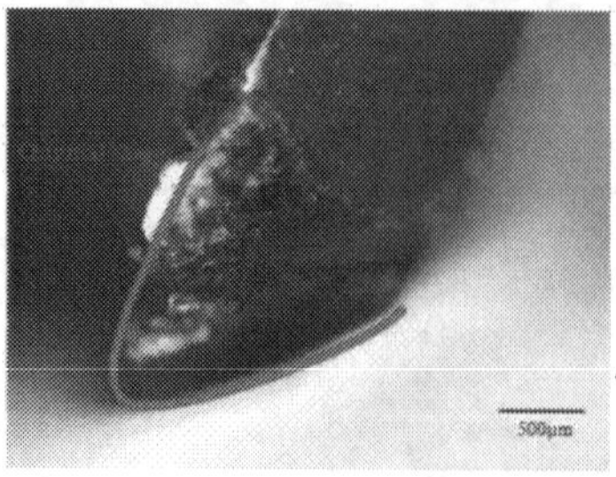

1280 Hz

Figure 10 Worn off cutting tool

3.2 Frequency domain analysis

Time domain signal can be converted into frequency domain signal by using the widely used method known as Fast Fourier Transform (FFT) to find out the content of the vibration response in frequency domain. In frequency domain, the vibration response often provides interesting information or the characteristic of the response. The vibration response from the milling process can be considered periodic time-varying and hence it is reasonable to apply frequency analysis method. Therefore, in this case, Fast Fourier Transform method is employed. The frequency domain analysis will present the effect of tool edge radius to the characteristic of the vibration response. In order to eliminate the static component of the vibration response, the static component of the response is filtered off before applying Fast Fourier Transform.

Fig. 3.6 shows the frequency spectra at different cutting speed for cutting tool No.1 and cutting tool No.2. At spindle speed of 1000 RPM, cutting tool No.1 shows numerous peak in the frequency band from 0 to 200 Hz, while for cutting tool No.2, there is a dominate peak at around 65Hz. Based on the spindle speed and number of flute of the cutting edge, the number of impacts during the milling process is computed at frequency of 16.67Hz. In other words, each of the cutting edge will hits on the workpiece for 16.67 times per second. Since the number of flutes are 4, the operating frequency is 66.68 Hz. Other operating frequencies are 133.36 Hz (2000 RPM), 200.04 Hz (3000 RPM), and 266.72 Hz (4000 RPM).

Magnitude values based on the operating frequencies are evaluated. From *Fig. 3.6 (a)-(d)*, it is observed that the magnitude are consistently lower (<400) with respect to the operating frequencies. *Fig. 3.6 (e)-(h)* shows otherwise, where as high magnitude (>1000) is observed at the operating frequency of 200.04 Hz. This observation coincides with the high RMS responses at 3000 RPM. The plot of FFT further justify the possibility of chatter occur at the 3000 RPM for Cutting tool No.2. However, it should be noted that as the operating frequency increases, the ploughing force also increases, and subsequently an increase in cutting temperature.

Therefore, it is deduced that with the increase of operating frequencies, an increase in the dynamic responses is expected, due to the increase of ploughing force. Unless a trigger towards the resonance frequencies is made, the dynamic responses will increase gradually as demonstrated by Cutting tool No.1. However, if the cutting operation trigger its resonance frequency, hence an extraordinary high vibration will occur at that frequency as demonstrated in Figure 3.6 (g) where a high magnitude coincides with its operating frequency. In this case, the tool edge radius is the main factor that causes the chatter vibration to occur.

On the other hand, *Fig. 3.6 (a)-(d)* also shows multiple peak at beyond the operating frequency, those peak occur may be due to the excessive noise that developed by cutting tool with lower tool edge radius. This phenomena also occur in cutting tool No.2, however, the magnitude value is much smaller compare to the cutting tool No.1. Under normal milling condition, the

https://doi.org/10.1051/matecconf/201820202006

magnitude value at operating frequency should be higher than the rest of the frequency band which also shown in the analysis of Huang [19]. For example, the magnitude value at operating frequency for *Fig 3.6 (a) and (e)* is higher than the rest of the frequency band. Therefore, it can be conclude that the milling process of cutting tool No.2 is more stable than cutting tool No.1 at 1000 RPM and 2000RPM with the assistance of process damping effect. However, as the cutting speed increases, the milling process of cutting tool No.2 at 4000 RPM is unstable due to higher cutting speed as mentioned in section 3.1, and multiple peak is clearly shown in *Fig 3.6 (h)*.

(a)

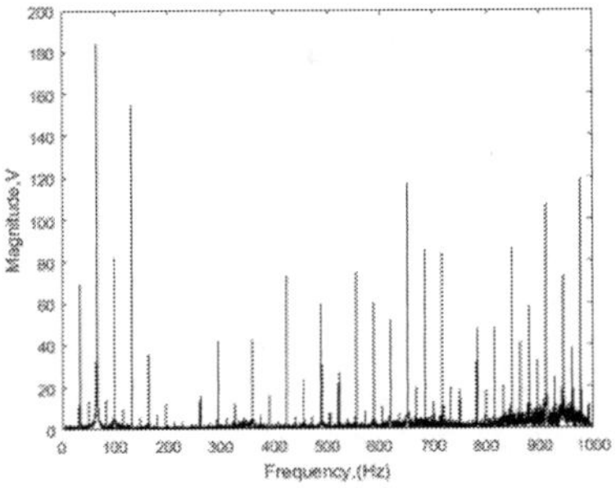

(b)

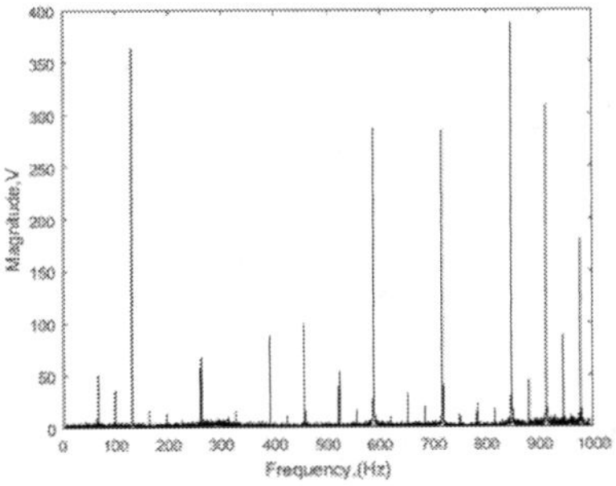

(c)

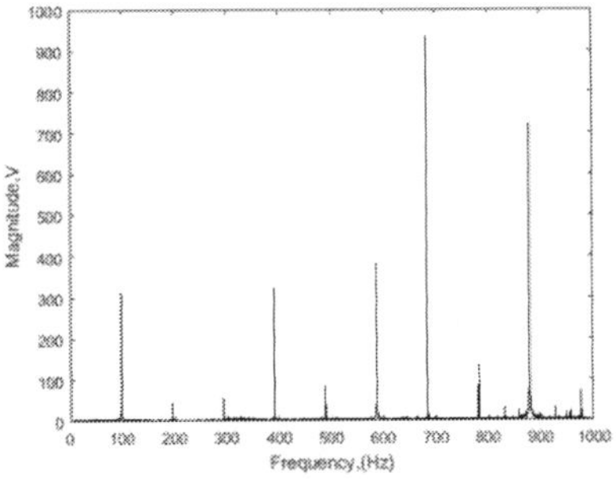

(d)

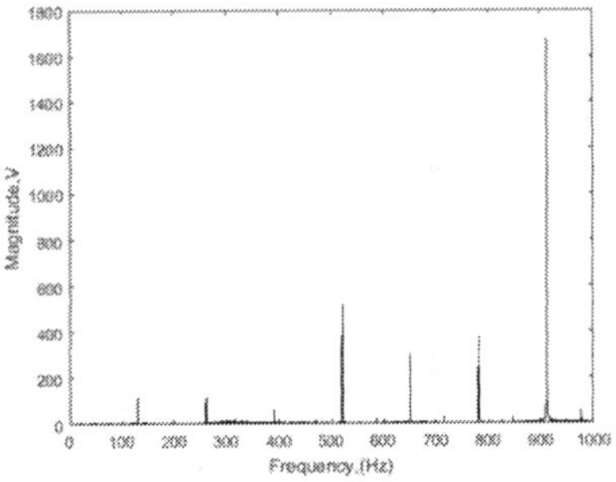

(e)

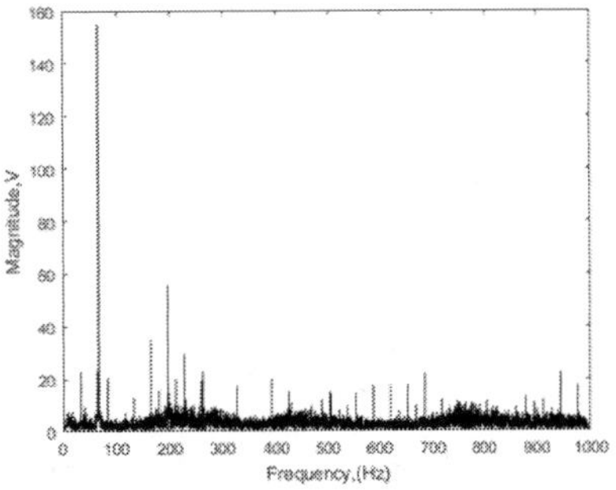

(f)

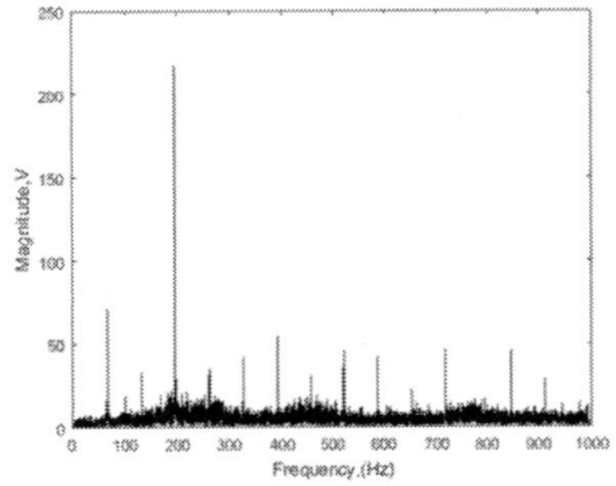

(g)

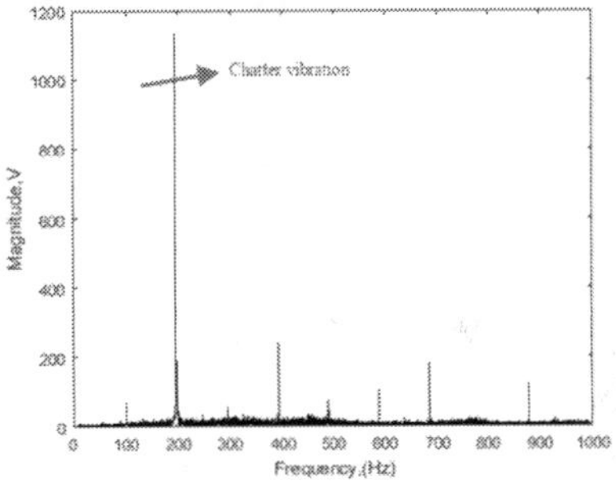

(h)

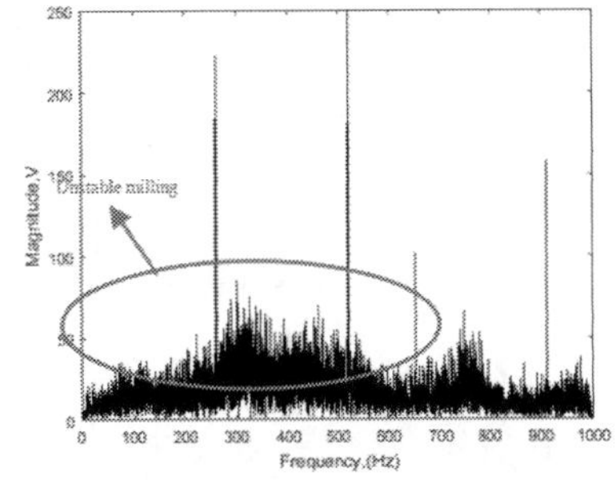

Figure 11 Frequency spectral for cutting tool No.1 (a) 1000 RPM, (b) 2000 RPM, (c) 3000 RPM, (d) 4000RPM, for cutting tool No.2 (e) 1000 RPM. (f) 2000 RPM, (g) 3000 RPM, (h) 4000 RPM

Acknowledgements

The authors are grateful to the Ministry of Higher Education (Malaysia) for providing the Fundamental Research Grant Scheme

(FRGS/1/2014/TK01/CURTIN/02/2) for this work to be possible.

References

[1] Ezugwu, E. O. 2005. "Key Improvements in the Machining of Difficult-to-Cut Aerospace Superalloys." International Journal of Machine Tools and Manufacture 45 (12–13): 1353-1367

[2] Dudzinski, D., A. Devillez, A. Moufki, D. Larrouquère, V. Zerrouki, and J. Vigneau. 2004. "A Review of Developments Towards Dry and High Speed Machining of Inconel 718 Alloy." International Journal of Machine Tools and Manufacture 44 (4): 439-456

[3] Ulutan, Durul, and Tugrul Ozel. 2011. "Machining Induced Surface Integrity in Titanium and Nickel Alloys: A Review." International Journal of Machine Tools and Manufacture 51 (3): 250-280

[4] Fang, Ning, P. Srinivasa Pai, and S. Mosquea. "Effect of tool edge wear on the cutting forces and vibrations in high-speed finish machining of Inconel 718: an experimental study and wavelet transform analysis." The International Journal of Advanced Manufacturing Technology 52, no. 1 (2011): 65-77.

[5] Pervaiz, Salman, Amir Rashid, Ibrahim Deiab, and Mihai Nicolescu. 2014. "Influence of Tool Materials on Machinability of Titanium- and Nickel-Based Alloys: A Review." Materials and Manufacturing Processes 29 (3): 219-252

[6] Jawaid, A., S. Koksal, and S. Sharif. 2001. "Cutting Performance and Wear Characteristics of Pvd Coated and Uncoated Carbide Tools in Face Milling Inconel 718 Aerospace Alloy." Journal of Materials Processing Technology 116 (1): 2-9

[7] Zhang, S., J. F. Li, and Y. W. Wang. 2012. "Tool Life and Cutting Forces in End Milling Inconel 718 under Dry and Minimum Quantity Cooling Lubrication Cutting Conditions." Journal of Cleaner Production 32: 81-87.

[8] Hadi, M. A., J. A. Ghani, C. H. Che Haron, and M. S. Kasim. 2013. "Comparison between up-Milling and Down-Milling Operations on Tool Wear in Milling Inconel 718." Procedia Engineering 68: 647-653

[9] Zhang, S., J. F. Li, and Y. W. Wang. 2012. "Tool Life and Cutting Forces in End Milling Inconel 718 under Dry and Minimum Quantity Cooling Lubrication Cutting Conditions." Journal of Cleaner Production 32: 81-87.

[10] Taylor, Fred W. "The Art of Cutting Metals." Scientific American 63 (1907): 25942-44.

[11] Wiercigroch, Marian, and Erhan Budak. "Sources of Nonlinearities, Chatter Generation and Suppression in Metal Cutting." Philosophical Transactions of the Royal Society of London A: Mathematical, Physical and Engineering Sciences 359, no. 1781 (2001): 663-93.

[12] Altintas, Yusuf. Manufacturing Automation: Metal Cutting Mechanics, Machine Tool Vibrations, and CNC Design. Cambridge university press, 2012.

[13] Quintana, Guillem, and Joaquim Ciurana. 2011. "Chatter in Machining Processes: A Review." International Journal of Machine Tools and Manufacture 51 (5): 363-376

[14] Tobias, SA, and W Fishwick. "Theory of Regenerative Machine Tool Chatter." The engineer 205, no. 7 (1958): 199-203.

[15] Sisson, Timothy Raymond, and Richard L Kegg. "An Explanation of Low-Speed Chatter Effects." Journal of Engineering for Industry 91, no. 4 (1969): 951-58.

[16] Slavicek, Jan. 1965. "The Effect of Irregular Tooth Pitch on Stability of Milling" Proc. of the 6th Int. MTDR

[17] Shirase (1996) Cutting force and dimensional surface error generation in peripheral milling with variable pitch helical end mills. Int J Mach Tool Manuf 36(5):567–584

[18] Altıntaş, Y., S. Engin, and E. Budak. "Analytical Stability Prediction and Design of Variable Pitch Cutters." Journal of Manufacturing Science and Engineering 121, no. 2 (1999): 173-78.

[19] Huang, Panling, Jianfeng Li, Jie Sun, and Maojie Ge. 2012. "Milling Force Vibration Analysis in High-Speed Milling Titanium Alloy Using Variable Pitch Angle Mill." The International Journal of Advanced Manufacturing

[20] Sellmeier, V, and B Denkena. "Stable Islands in the Stability Chart of Milling Processes Due to Unequal Tooth Pitch." International Journal of Machine Tools and Manufacture 51, no. 2 (2011): 152-64.

[21] Comak, Alptunc, and Erhan Budak. "Modeling Dynamics and Stability of Variable Pitch and Helix Milling Tools for Development of a Design Method to Maximize Chatter Stability." Precision Engineering 47 (2017): 459-68.

[22] Denkena, B., and D. Biermann. "Cutting Edge Geometries." CIRP Annals - Manufacturing Technology 63, no. 2 (2014): 631-53.

[23] Endres, William J., and Raja K. Kountanya. "The Effects of Corner Radius and Edge Radius on Tool Flank Wear." Journal of Manufacturing Processes 4, no. 2 (2002/01/01/ 2002): 89-96.

[24] Gurdal, Ozan, Erdem Ozturk, and Neil D. Sims. "Analysis of Process Damping in Milling." Procedia CIRP 55, no. Supplement C (2016/01/01/ 2016): 152-57.

[25] Yusoff, AhmadR, Sam Turner, ChrisM Taylor, and NeilD Sims. "The Role of Tool Geometry in Process Damped Milling." The International Journal of Advanced Manufacturing Technology 50, no. 9-12 (2010/10/01 2010): 883-95.

[26] Albrecht, Paul. "New Developments in the Theory of the Metal-Cutting Process: Part I. The Ploughing Process in Metal Cutting." Journal of engineering for industry 82, no. 4 (1960): 348-57.

[27] Wyen, C-F, and Konrad Wegener. "Influence of Cutting Edge Radius on Cutting Forces in Machining Titanium." CIRP annals 59, no. 1 (2010): 93-96.

Analysis on Rolling Damping of a Conventional Boat fitted with T-shaped Bilge Keels

Yam Ke San, Gordon Chiew Chin Howe, Vincent Lee Chieng Chen, and Sukanta Roy

Department of Mechanical Engineering, Curtin University Malaysia, CDT 250 Miri Sarawak Malaysia.

Abstract. This work presents a numerical study on the effect of T-shaped bilge keels on the roll damping of a conventional boat. A scaled boat model with the same dimensions as that of Irkal et al. [2] was fitted with two T-shaped bilge keels at the edges of the model. Computational Fluid Dynamics method was employed to simulate the roll decay motion of the boat. The motion of the boat is captured using a 6DOF model and the Overset grid approach. Comparison was performed on the damping characteristics of the conventional I-shaped and the T-shaped bilge keels. In addition, the impact of the aspect ratio of the keel bilges on the roll damping of the boat was evaluated. It was found that the bilge keel aspect ratio influences the damping coefficient non-linearly. Sufficiently large aspect ratio, i.e. an aspect ratio greater than 2, is necessary in order to obtain an effective damping on the peak angle.

1 Introduction

Stability is the ability of a boat vessel to return to its original position due to the presences of centre of buoyancy and vertical centre of gravity. When the negative stability occurs, boat will capsize due to the unbalance forces acting at the vessels. Thus the heave, pitch and roll are important characteristics for determining the stability of ships [1]. For rolling, the use of bilge keels is the most common method for damping the rolling motion. In the past, theoretical study on bilge keels designs is difficult owning to the significant viscous and eddies effects from the hull on the roll damping. Experimental method is more common. However, the advancement of computational fluid dynamic (CFD) has enabled theoretical study to simulate the complex viscous flows around the hull which significantly enhances the accuracy of the prediction on the roll damping. Indeed, a recent study by Irkal et al. [2] on the decay rolling motion of a barge with bilge keels has shown that a good match between the numerical method prediction and their experimental result. A comparison study on the full numerical method of a boat with bilge keels against the traditional inviscid result [3] also have shown that the inviscid approach significantly under-predicted the damping at high wave frequency range.

The shape and width of the keel bilges has been known to affect the damping coefficient of a boat [4-12]. The more vortices a bilge keel can generate the higher roll damping it can achieve [5]. A comparison study on round and sharp bilge keels [6] shows that sharp bilge keel is able to generate stronger vortices at the tip of the bilge keels and therefore has a greater damping coefficient. Using complex shape [7] has also shown to be able to generate more vortices and increase the roll damping.

This paper addresses the use of T-shaped keel bilges on roll damping of a common barge. T-shaped is a slight extension of the common I-shaped bilge keel. However, the increase of the surface area is expected to be able to induce strong vortices that can enhance the roll damping. In order to evaluate the effectiveness of the T-shaped bilge keels, comparison against the I-shaped will also be performed.

2 Methodology

We consider the scaled boat model used by Irkal et al. [2] as the model of the investigation. The model has a two T-shaped keel bilges fitted at both of its lower edges and is partially immersed in a tank as shown in Fig. 1. The boat is locked at its centre of mass, and is initially resting at an inclination angle $\emptyset_0$ of 20° at t = 0s, and is then released from rest so that it can rotate about its transverse axis, as displayed in Fig. 1. The dimensions of the boat are 300 mm in breadth (i.e. beam, B) and 200 mm in depth (D) so that the beam to depth ratio (B/D) is 1.5. The width of the boat is 580mm (into the paper). The length of the flume of the tank is 10 times the length of the boat, the height is 800mm and the width is 600mm. The water in the tank is filled up to half the height of the tank, and the boat is allowed to immersed in the water such that the ratio of the depth of the water tank to the draft of the boat is 1.67. The total displacement of a water mass by the boat is 20.88kg. The boat is made out of Perspex sheet, and possesses a moment of inertia I value of 0.2248 kg.m^2.

The first objective is to measure the free decay rolling of boat that is fitted with T-shaped bilge keels and

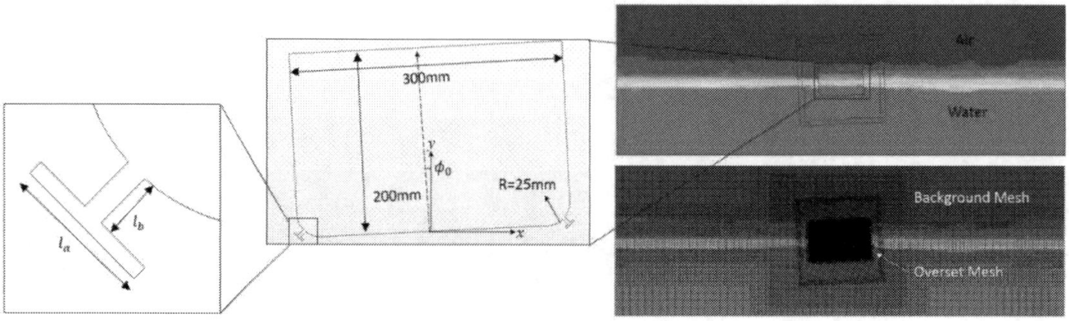

Figure 1 (*left*) length of the arm and body of a T-shaped bilge keel, (*middle*) the dimensions of a boat with T-shaped bilge keel, (*top right*) air and water domain, (*bottom right*) mesh distribution in the background and the Overset volume.

compare against that of conventional I-shaped bilge keels. The second objective is to investigate the effect of increases the aspect ratio of the T-shaped keel bilge, where the aspect ratio *AR* of T-shaped is defined as the ratio of the length of the arms l_a and the body l_b as shown in Fig. 1.

Commercial Computational Fluid Dynamics (CFD) code StarCCM is employed to simulate the problem. Volume of Fluid (VOF) method is employed to simulate the air and water phases and the free surface interface. The motion of the boat is calculated using the Six Degree of Freedom (6DOF) model. However, in this case, the model is reduced to only 1DOF since the boat experiences the rotation only in one axis. The interaction between the motion of the boat and the hydrodynamic of the fluid is captured via force exchanges at the boundary boat. The fluid is assumed to be in laminar c ondition at atmospheric pressure. The surface boundary of the boat and the tank are assumed to be a wall (zero velocity at the surface). Because the front and the back of the boat are the same, we consider only half of the actual boat and tank in order to save computational cost. Symmetry boundary condition is then applied to the surface at the half of the domain.

Structured mesh is deployed in the computational domain for computing the fluid pressure and velocity. High mesh density is being applied in the vicinity of the water surface in order to capture a sharp transition of water free surface. Also high mesh density is applied at the boat boundary in order to capture the complex flows surrounding the bilge keels and to increase the accuracy of the force interaction calculation. A separate domain of structured mesh that overlaps with the background mesh (called the Overset mesh) is created for the boat to accommodate the relative motion of the boat. The coupling between the two meshes is performed via grid to grid connectivity model developed by StarCCM.

The governing equations are solved using Finite Volume Method. Second-order accurate discretization schemes are selected to achieve accurate solutions. The linear second-order upwind scheme was selected to discretize the spatial advection terms. First order Euler implicit scheme with a time step of 0.05s was selected to numerically advance the time of the solution. The conservation of mass and momentum are enforced through segregated solver where the pressure, velocity and the phase volume fraction equations are solved sequentially and iteratively until the criterion on the minimum residue (a value of 1e-4) is met for each equation. In addition, a sharpening factor of 1 is applied to the VOF model to enhance the water-air interface.

A total 7 simulations were performed. One is on I-shaped bilge keel, and the remaining are on T-shaped bilge keels with the aspect ratio *AR* ranges from 0.5 to 3.

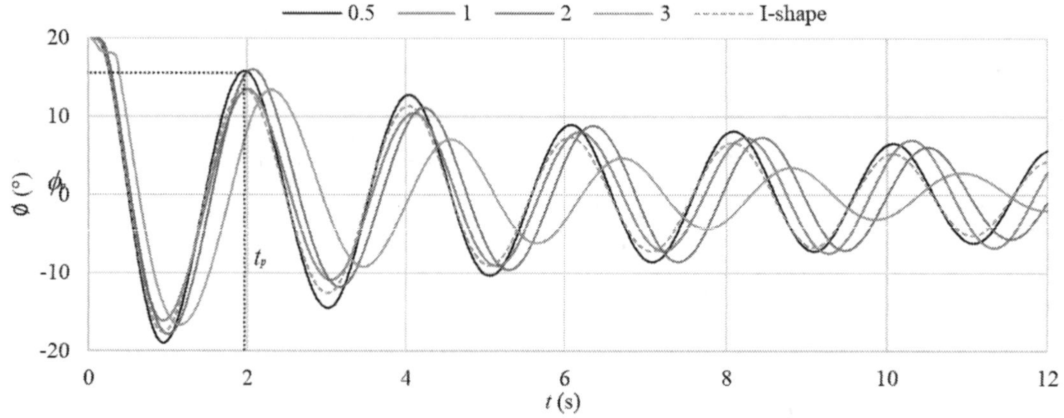

Figure 2 Roll angle history for boat with I-shaped bilge keels or T-shaped of different aspect ratio

https://doi.org/10.1051/matecconf/201820202007

Each simulation is performed up to a period of 20secs, but the analysis focusses on the first five cycles where damping is essential.

3 Result & discussion

3.1 Period and peak angle

Figure 2 shows the roll angle history of the boat with T-shaped bilge keels of aspect ratios ranged from 0.5 to 3, for a period of 12s. All designs exhibit similar sine wave profile with a decaying amplitude over the oscillations. The key differences are the rate of the decay of the roll amplitude and also the timing of the cycle's peak angle t_P. It is clear from the figure that the lowest aspect ratio AR = 0.5 shows the largest amplitude in almost every cycle, which implies it has the lowest decrease of the amplitude. On the other hand, the largest aspect ratio AR = 3 exhibits the lowest amplitude and hence shows the greatest decrease amplitude, over the oscillations. For AR = 1 and 2, there is no significant decrease of the amplitude. As for the timing of the peak angle t_p, AR = 0.5 shows the earliest time, whereas AR = 3 clearly shows the latest, indicating the cycle period is increased for larger aspect ratio. For the aspect ratio of 2 and 3, the period also increases over the cycles. It can be concluded that increasing the aspect ratio increases the damping of the rate of rolling but only increases the magnitude of the roll damping when the aspect ratio is increased to 3.

Also shown on Figure 2 is the roll angle history of the boat with I-shaped bilge keels. The profile is quite similar with the AR = 0.5, i.e. same timing of the peak angle t_p but slightly lower amplitude. As a summary, the peak roll angle ϕ_p from Fig. 2 is replotted in Fig. 3, with a further addition of two interval aspect ratios AR = 1.5 and 2.5. All cases show a 4th order polynomial decrease. However, we also can clearly observe multiple crosses between the curves, indicating a non-linear relationship between the roll angle and the aspect ratio.

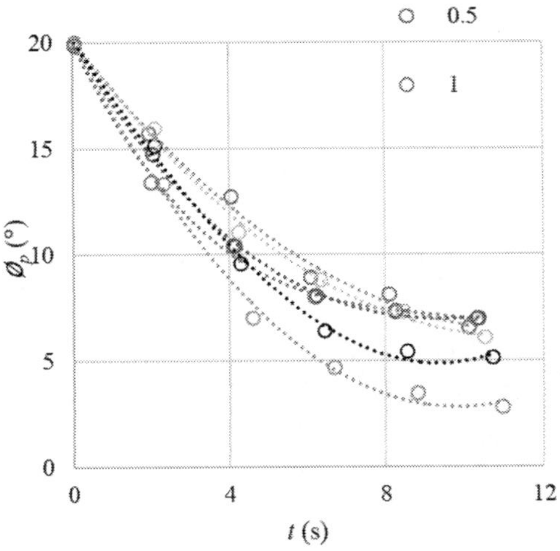

Figure 3. Peak roll angle ϕ_p as a function of t for different aspect ratio AR.

The effect of changing the aspect ratio on the can be further quantified by considering the increment of the period t_a over the cycles, with respect to the period for case AR = 0.5, as shown in Fig. 4. Increasing the AR almost always increases the period of the cycle, and the trend is an exponential increase with an increasing power for the later cycles. For example, at the first cycle, the aspect ratio AR = 3 increases the period by 18% compared to only 8% and 2% for AR = 2 and 1, respectively. But at the fifth cycle, the period is significantly increased by 42% compared to 20% and 10% for AR = 2 and 1, respectively.

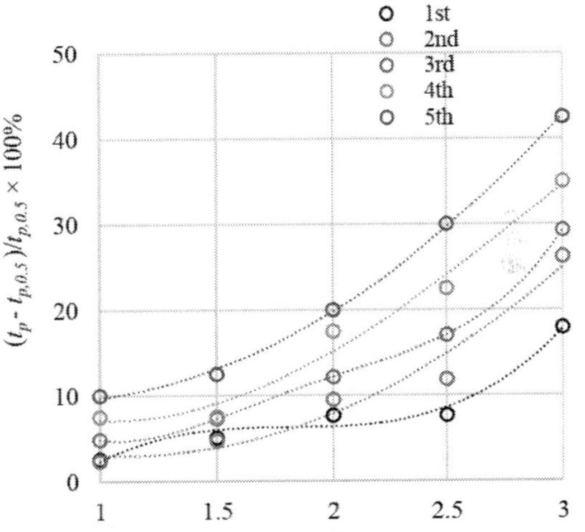

Figure 4. The percentage of period increment as a function of AR for the first five cycles, with respect to the period of case AR = 0.5, $t_{p,0.5}$.

The effect of the aspect ratio on the peak angle ϕ_p was

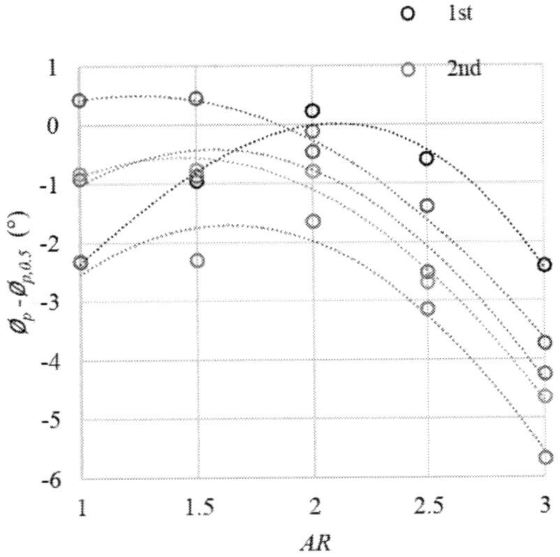

Figure 5. The change of peak roll angle ϕ_p with respect to the peak roll angle of case AR = 0.5, $\emptyset_{p,0.5}$, as a function of AR for the first five cycles

also further analysed. The change of the peak angle ϕ_p with respect to the peak angle of case AR = 0.5 is shown in Fig. 5. It can be seen that increasing the aspect ratio of the bilge keels influences the amplitude non-linearly. The first cycle exhibits a quadratic curve but it gradually transitions into an exponential decrease at the fifth cycle. For the first four cycles, the amplitude increases with the increase of AR and only decreases when AR is increased beyond 2. Only in the last cycle, the roll angle always decreases for any increase of AR. It can be concluded that aspect ratio greater than 2 is necessary for an effective decrease on the roll angle. The maximum decrease of the roll angle varies from cycle to cycle. The peak angle is reduced by a maximum of only 2 degrees in the first cycle. It is 6 degrees in the second cycle, and between 2 and 6 degrees for the third to fifth cycles.

3.2 Damping coefficient

It is customary to describe a free roll decay motion using a linear equation of motion

$$(I + A)\ddot{\phi} + B_l\dot{\phi} + C\phi = 0 \quad (1)$$

where $(I + A)$ is the roll moment of inertia, B_l is the damping coefficient and C is the stiffness coefficient. The linear damping coefficient is given by

$$B_l = 2(I + A)\zeta\omega_n \quad (2)$$

The damping ratio ξ and natural frequency ω_n are defined as

$$\xi = \frac{1}{\sqrt{1+\left(\frac{2\pi}{\delta}\right)^2}} \quad (3)$$

$$\omega_n = \frac{2\pi}{T_n} \quad (4)$$

where δ is the logarithmic decrement of two successive peaks. Based on the first six successive peaks, the linear damping coefficient is then computed using equations (2) to (4). The result is shown in Figure 6. It can be seen that the behavior is a quadratic trend. The damping initially decreases and then increases when AR is greater than 1.5. This again shows a nonlinear effect of the bilge keel size on the roll moment and pressure around the hull, in agreement with Irkal et al. [2] observation.

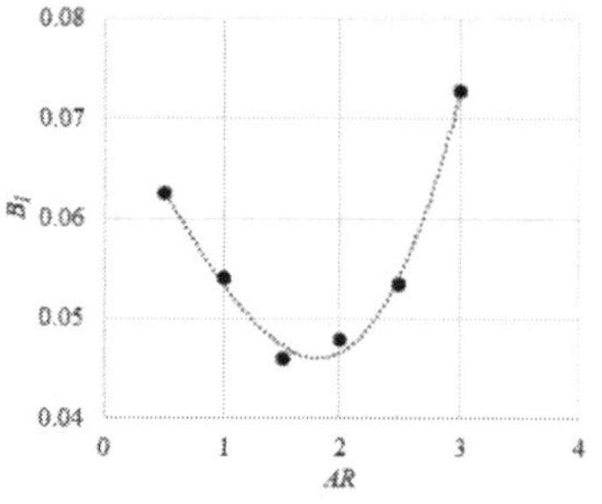

Figure 6. B_l as a function of AR.

3.3 Eddies and vortices

The underlying physics that leads to the increase of the damping is the influence of the bilge keels on the flow surrounding the boat. The keels obstruct the flow close to the boat and generates vortices that increases the pressure and the shear force on the boat. This mechanism is shown on Figure 5 for the I-shaped, and T-shaped keel of aspect ratio AR = 1 and 3. However, it can be seen that T-shaped keel is able to generate significantly greater vorticity and eddies at the tip of the bilge keel, a maximum value of $200s^{-1}$, which is four times greater than that of either I-shaped keel or T-shaped of AR = 1. The large vorticity increases the pressure being applied at the tip of bilge keel. This highlights the advantage of increasing the surface area of the bilge keel. However, AR greater than 2 is necessary in order to achieve vortices generation that is significantly greater than that of I-shaped keel bilge.

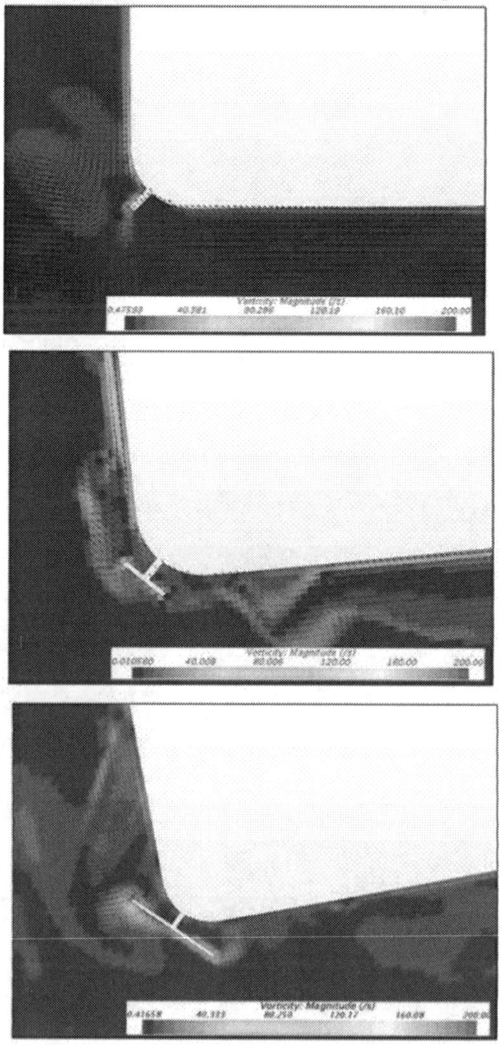

Figure 7. Vorticity magnitude and velocity vector contour of (*top*) I-shaped, (*middle*) T-shaped with AR = 1, (*bottom*) T-shaped with AR = 3, at time t = 2.4s.

4 Conclusions

This work presents a numerical study on the effect of T-shaped bilge keels on the roll damping of a conventional boat. The analysis focusses on the effect of the aspect ratio of the T-shaped bilge keels on the peak roll angle and the period of the oscillations. The following conclusions are drawn:

1. Increasing the bilge keel aspect ratio influences the damping coefficient non-linearly. Due to this effect, the peak roll angle also changes non-linearly. As a result, sufficiently large aspect ratio, i.e. aspect ratio greater than 2, is necessary in order to obtain an effective decrease of the peak angle.
2. The damping on the period of the cycle, on the other hand, increases consistently and exponentially with an increasing aspect ratio. The effect is greatly increased at later cycle.
3. The non-linear damping is attributed to the non-linear increase of the eddies and vortices generation at the tip of the keel.

References

1. R.A. Ibrahim, I.M. Grace. Modeling of ship roll dynamics and its coupling with heave and pitch. Mathematical Problems in Engineering, 2010.
2. M.A. Irkal, , S. Nallayarasu, S.K. Bhattacharyya. CFD approach to roll damping of ship with bilge keel with experimental validation. Applied Ocean Research, **55**, pp.1-17. 2016.
3. E.P. Bangun, EP, C.M. Wang, T. Utsunomiya. Hydrodynamic Forces on a Rolling Barge with Bilge Keels." Applied Ocean Research, **32(2)**, 219-232. 2010
4. Y. Ikeda. Prediction methods of roll damping of ships and their application to determine optimum stabilization devices. Marine Technology, **41(2)**, pp.89-93. 2004
5. K.P. Thiagarajan, E.C. Braddock. Influence of Bilge Keel Width on the Roll Damping of Fpso. Journal of Offshore Mechanics and Arctic Engineering, **132(1)**, 011303. 2010
6. F. Jaouen, A. Koop, G. Vaz. Predicting roll added mass and damping of a ship hull section using CFD. In ASME 30th International Conference on Ocean, Offshore and Arctic Engineering, pp. 105-115. 2011
7. R. van't Veer, X.B. Schut, R.H.M. Huijsmans. Bilge Keel Loads and Hull Pressures Created by Bilge Keels Fitted to a Rotating Cylinder. Applied Ocean Research ,**53**, 1-14. 2015
8. S.A. Kinnas, Y.H. Yu, V. Vinayan. Numerical methods for the prediction of the bilge keel effects on the response of ship-shaped hulls. In ASME 26th International Conference on Offshore Mechanics and Arctic Engineering, pp. 1023-1030. 2007
9. Y.J. Kim, I.K. Kang, B.S. Park, S.J. Ham. An optimal bilge keel design to reduce the rolling of the offshore large purse seiner. Journal of the Korean society of Fisheries Technology, **50(2)**, pp.147-153. 2014
10. B. Ommani, T. Kristiansen, O.M. Faltinsen, Simplified CFD modeling for bilge keel force and hull pressure distribution on a rotating cylinder. Applied Ocean Research, **58**, pp.253-265. 2016
11. G.O.G. Avalos, J.B. Wanderley. Numerical study of forced roll oscillation of FPSO with bilge keel. Ocean Engineering, **147**, pp.304-317. 2018
12. K.G. Aarsæther, D. Kristiansen, B. Su and C, Lugni. Modelling of Roll Damping Effects for a Fishing Vessel with Forward Speed. In ASME 34th International Conference on Ocean, Offshore and Arctic Engineering . 2015

Performance Comparison of Organic Rankine Cycle and Vapour Compression Cycle Hybrid Cooling-Heating System using Subcritical or Supercritical Working Fluid

Yam Ke San , Raja Wahiduzzaman Bin Raja Ismail , Vincent Lee Chieng Chen , and Jung Hyung Chul

Department of Mechanical Engineering, Curtin University Malaysia, CDT 250 Miri Sarawak Malaysia.

Abstract. This paper presents a mathematical modelling on the evaluation of cooling, heating and power performance of a hybrid system of Organic Rankine Cycle and Vapour Compression Cycle. The system is assumed to be powered through solar parabolic trough collector and is able to generate a cooling power of 10 kW. Refrigerants R134a or R245fa are chosen as the working fluid of the system. The system is constructed using commercial energy modelling tool AspenPlus. Analysis is performed to determine the effect of changing the mass flow rate split ratio on the energy output. The effect of using subcritical and supercritical working fluid is also compared. Particular attention is paid toward the condition where the power output is equivalent to the energy consumption in view of creating a self-powered cooling and heating system. The result shows that the coefficient of performance for system using R245fa is higher compared to that using R134a. However, the systen using R134a allows a self-powered cooling and heating system to be achieved to be achieve at a much higher mass split ratio, resulting the system to be 35% more efficient in the performance.

1 Introduction

Air conditioning contributes to approximately 15% of the total energy consumption in Malaysia [1], that mainly comes from coal-fueled power plants with negative carbon footprints. As a measure to curb the consumption of non-renewable and carbon emission, this study looks at the feasibility of the Organic Rankine Cycle-Vapor Compression Cycle (ORC-VCC) system as a green energy powered hybrid system that provide air cooling and water heating for household application.

The use of Organic Rankine Cycle to harvest power from low-grade heat source is not new and has been known to be a viable technology for converting heat to electrical power in various of the system that ranges from micro-scale vehicle exhaust system [2] to large-scale 30MW of geothermal power plant [3]. Increasing interest in renewable energy worldwide has prompted significant research and development in the field. One drawback of the system that researchers have been addressing is its low efficiency. Research have shown that the efficiency can be significantly improved by altering or revamping the cycle architectures [4,5], or by optimizing the selection of working fluid [6,7]. Another increasingly popular approach to improve the efficiency is through cogenerations. Cogeneration minimizes the system size by combining the ORC with other energy system and converts all form of energy exchanges within the cycles into useful output. Some examples of cogeneration unit include combining the waste heat from biomass combustion and the solar power to increase the critical temperature [8], integrating the solar-powered ORC to a desalination process [9,10], integrating ORC into solid fuel cell and micro-gas turbine gasification [11], and combining ORC to a Vapor Compression Cycle (VCC) to produce both cooling and heating [12].

The research on ORC-VCC is still relatively new. A recent study [12] evaluated the feasibility of a combined biomass and solar powered ORC-VCC in providing cooling and water heating in Greece during the summer season and concluded the feasibility of the system. However, the study uses the subcritical working fluid and did not report the effect of the mass flow rate ratio. Understanding the effect of the mass flow rate ratio on the energy balance is critical for designing an optimum system. In addition, the selection of the working fluid condition also is known to be pivotal in the system efficiency [4]. This study evaluates the effect of the mass flow rate ratio and the performance of the subcritical and supercritical working fluid R245fa in a solar-powered ORC-VCC.

2 Methodology

2.1 System description

We consider a ORC-VCC system consists of a solar-powered ORC that is interconnected to the VCC system. The ORC system consists of a solar heater, an expander, a pump and a condenser, whereas the VCC system consist of an evaporator, a compressor, a condenser, and an expansion valve, as shown in Fig. 1. Both systems share the same condenser, which means the flows from both systems joins together to enter the condenser, and then are split upon exiting the condenser. The primary function of the ORC system is to convert the solar heat into power, that is used to drive the motor of the compressor in the VCC system. In this case compressor is the only component that consumes significant power. Pump typically requires very little work to operate and its energy consumption can be neglected. The VCC system functions as an air cooler, which absorbs the heat through the evaporator. The thermodynamic system is closed by having the condenser as a heat remover. The heat taken out from the condenser is utilized for heating water. Therefore, the whole system operates as a cogeneration unit (cooling and heating). The detail of the flow sequences is listed in Table 1.

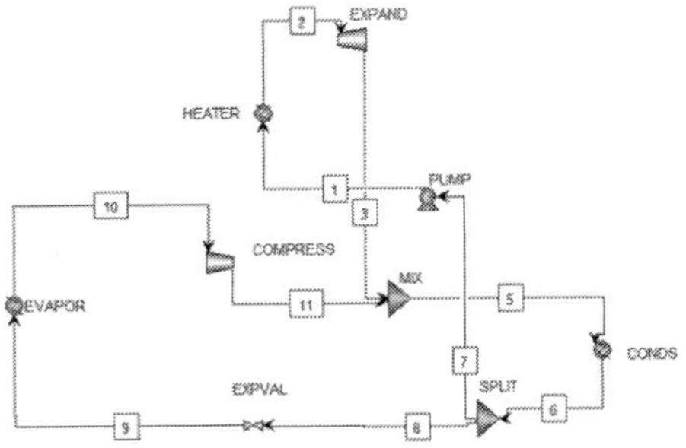

Figure 1. Components of the ORC-VCC system developed in Aspen Plus.

Table 1. The state point of the ORC-VCC.

State point	Description
1-2	Inlet and outlet of the solar PTC
2-3	Inlet and outlet of the turbine expander
{3,11}-5	Inlet ad outlet of the mixture
5-6	Inlet and outlet of the condenser
6-{7,8}	Inlet and outlet of the splitter
7-1	Inlet and outlet of the pump
8-9	Inlet and outlet of the expansion valve
9-10	Inlet and outlet of the evaporator
10-11	Inlet and outlet of the compressor

2.2 System constraints

The key operational parameters are the temperature and the pressure of the working fluid at the inlet of the ORC expander, and the condensation temperature/pressure. For direct vapour generation saturation operation, the maximum operating temperature of a Parabolic Trough Collector (PTC) is 250 °C [9]. The pressure at the heater is set at 3MPa for subcritical and 6MPa for supercritical fluid. The condensation temperature for the system is set at 80°C, in order to cover domestic hot water demands. The evaporation temperature of the VCC is set equal to 10°C, which is reasonably low for air cooling purpose. The total mass flow rate of the system $\dot{m}_t$ is set at 0.32kg/s. Finally, the cooling load has to be specified but the value depends on the applications. Many factors such as the weather, and the building characteristics (size, material etc) can influence the cooling demand of a building. A heating load of 10kW is specified to the evaporator. The average Malaysia electricity consumption is about 1kW [13]. Therefore, the system can provide cooling up to ten households. In order for the system to be able to generate sufficient cooling power without relying on the grid electricity, the ORC must generate sufficient electricity power for the operation of the compressor and the pump. This requires a balance of the mass flow rate in the ORC and the VCC cycle. Therefore, this work aims to analyze the effect of changing the balance of the mass flow rate in the system, i.e. the ratio of the mass flow rate into the VCC $\dot{m}_c$ to the total mass flow rate $\dot{m}_t$,

$$R_m = \frac{\dot{m}_c}{\dot{m}_t} \quad (1)$$

, and also to compare the performance of the cycle using subcritical and supercritical fluid. Equation (1) is termed as mass split ratio.

The selection of suitable working fluid is essential to ensure high thermal efficiency and high utilization of the available heat source. Two commonly employed refrigerants, i.e. R134a and R245fa are chosen as the working fluid for the organic Rankine cycles given its desirable thermodynamic and thermal transport properties such as low boiling temperature and relatively high thermal conductivity, which makes it an ideal for low to medium temperature applications. The dry nature of the fluid ensure that the fluid remains in superheated state at the end of the expansion. In addition, it is nonflammable, has very low toxicity, non-corrosive and is environmentally friendly. The properties of both fluids are listed in Table 2.

Table 2 Physical, safety and environmental data of the working fluids used this [6]

Properties	R134a	R245fa
Molecular mass (kg/kmol)	102.03	134.05
Critical Temperature (K)	374	427.2
Critical Pressure (K)	4.059	3.64
Atmospheric lifetime	14.0	7.2
Global Warming Potential	1430	900
Ozone depletion potential	0	0

 https://doi.org/10.1051/matecconf/201820202008

2.3 Mathematical modelling

The performance of the system is evaluated using first and second law of thermodynamics. The system is assumed to operate under steady-state assumption. For simplicity, pressure drops and heat losses in the equipment and the pipes are neglected. The total heating generated from a solar collector can be expressed as

$$Q_s = \dot{m}_p(h_1 - h_2) \quad (2)$$

where $\dot{m}_p$ represents mass flow rate of working fluid in the ORC cycle, and the power generation at turbine expander is

$$W_t = \dot{m}_p(h_2 - h_3)\eta_t\eta_m \quad (3)$$

where η_t is the isentropic efficiency and η_m is the mechanical efficiency for the expander. The energy consumption of the pump is

$$W_p = \dot{m}_p(h_7 - h_8)\eta_p \quad (4)$$

The amount of heat energy absorbed by the evaporator of the VCC is

$$Q_e = \dot{m}_c(h_{10} - h_{11}) \quad (5)$$

The energy consumption of the compressor is

$$W_c = \dot{m}_c(h_{11} - h_{12})\,\eta_c \quad (6)$$

The total energy consumption of the system therefore is $W_T = W_c + W_p$. The heat removal by the condenser is

$$Q_c = \dot{m}_t(h_5 - h_6) \quad (7)$$

where $\dot{m}_t = \dot{m}_p + \dot{m}_c$. The coefficient of performance (COP) for the VCC is

$$COP_{VCC} = \frac{Q_c}{W_c} \quad (8)$$

The net COP for the system is

$$COP_{Net} = \frac{Q_e+Q_c}{Q_s} \quad (9)$$

The efficiency of the ORC is

$$\eta_p = \frac{W_t}{Q_s} \quad (10)$$

The ratio of the turbine power output to the energy consumption is

$$R_w = \frac{W_t}{W_c+W_p} \quad (11)$$

The ratio of the cooling and heating energy is

$$R_h = \frac{Q_e}{Q_c} \quad (12)$$

The value for the efficiency of the system components is listed in Table 3. The system is constructed in a commercial energy simulator AspenPlus, in which the thermodynamic calculation is performed.

Table 3 Efficiency of the system component [14].

Assumption	Value
Pump isentropic efficiency η_p (%)	65
Expander isentropic efficiency η_t (%)	60
Compressor isentropic efficiency η_c (%)	75
Mechanical efficiency η_m (%)	98

3 Result & discussion

Figure 2 shows the P-h diagram for the ORC-VCC cycle. Using supercritical fluid increases the area of the ORC system, which increases the expander output. By doubling the pressure in the boiler, the expander output is tripled. The effect of using supercritical on the temperature-entropy diagram is also shown in Figure 2. The area of the ORC is doubled, which allows the fluid to reach the maximum temperature (250 °C) at a lower value of entropy and increases the expander power output.

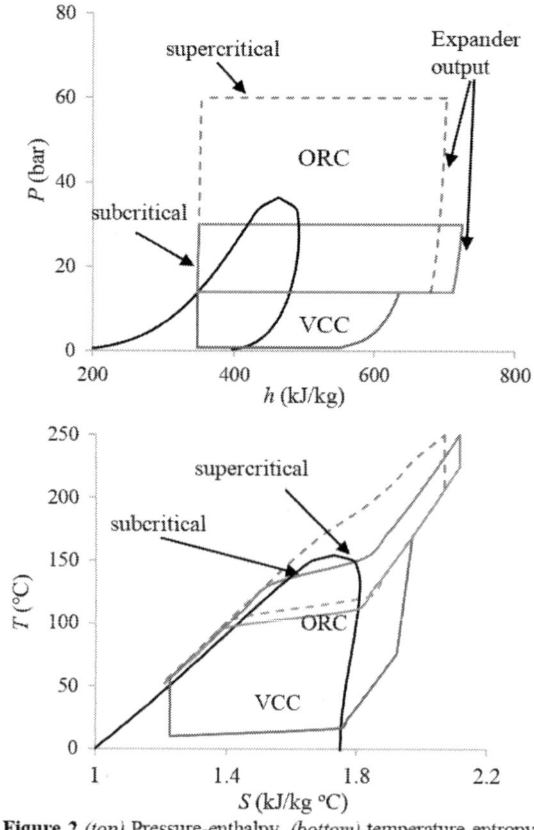

Figure 2 *(top)* Pressure-enthalpy, *(bottom)* temperature-entropy diagram for the ORC-VCC system using R245fa.

https://doi.org/10.1051/matecconf/201820202008

Figure 3 shows the effect of the mass flow rate split ratio on the total work consumption, and the expander power output for both subcritical and supercritical cycle. Increasing the flow rate into the VCC system linearly increases the consumption, whereas the lower flow rate into the ORC cycle linearly decreases the power output. It can be seen that employing supercritical cycle doubled the turbine output at R_m = 0.1. However, the rate of decrease of the turbine energy also increases for supercritical cycle, which decreases its difference from the subcritical case at large R_m. On other hand, there is no significant changes to the consumption energy for both cycles.

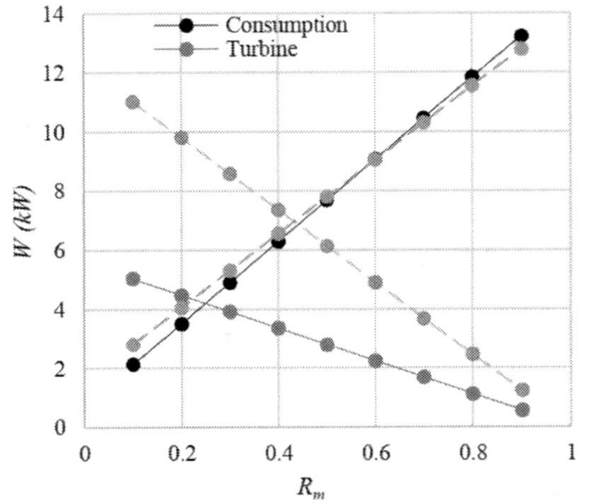

Figure 3. W_T and W_t as function of R_m for subcritical (*bold lines*) and supercritical cycle (*dashed lines*) using R134a.

In order to design a self-powered 10kW cooling system, we seek for the flow rate split ratio R_m that could yield a cooling load of 10kW at the ratio of the turbine power output to the energy consumption $R_w \geq 1$. When $R_w = 1$, the system produces a power output equal to its consumption, and an ideal maximum cooling power. When $R_w \geq 1$, the system produces cooling less than its ideal maximum, but also a surplus of electricity. When $R_w < 1$, the system produces a cooling load greater than its ideal maximum but requires additional power from external source in order to operate. Clearly it is desirable to achieve the work ratio $R_w = 1$ at a large value of the flow rate split ratio R_m since it improves the system efficiency and provides a greater turbine output. This can be seen in Fig. 4 which illustrates the relationship between R_w and R_m for both refrigerants using subcritical or supercritical cycles. In all cases, R_w decreases exponentially against R_m, for both subcritical and supercritical fluid. Also all have very similar gradients, in the range of $e^{-4.3Rm}$ to $e^{-4.8Rm}$. Since the gradient of both lines are approximately the same, the difference between subcritical and supercritical can be inferred by subtracting the equation of both lines. Therefore, the supercritical cycle is consistently offset the subcritical by a constant value, giving supercritical cycle having a greater value of R_m at $R_w = 1$. For system using R245fa, supercritical cycle increases the value of R_m from 0.1 to 0.2. For system using R124a, supercritical cycle greatly increases the value of R_m from 0.28 to 0.42.

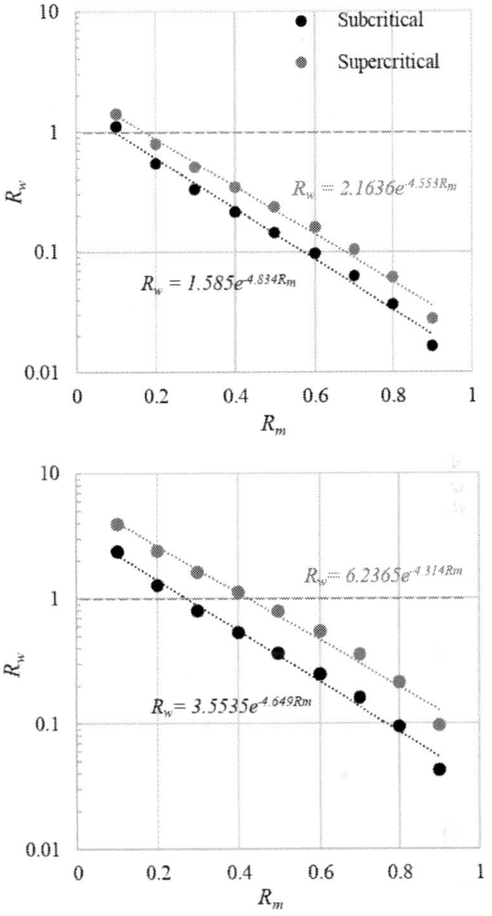

Figure 4. R_w as function of R_m for system using R245fa (*top*) and R134a (*bottom*). The blue dashed line denotes R_w = 1.

The efficiency of the system is evaluated using the coefficient of performance (COP). The gross COP computes the efficiency in the conversion of the heat source to the total useful output (both cooling and heating $Q_c + Q_e$). As shown in Figure 5, the COP increases exponentially with the increase of R_m for both refrigerants in either subcritical or supercritical cycle, i.e. the system is highly efficient at very large R_m. The use of supercritical increases the COP slightly. Also, in overall, R245fa produces slightly a higher COP than R134a. However, for the case R_w = 1, the low value of R_m for R245fa results in the low COP. The COP of supercritical and subcritical for R245fa at R_w = 1 is 1.3 and 1.1, respectively. Therefore, the supercritical cycle increases the COP by 0.2, which is about 18% higher than subcritical cycle. On the other hand, the high value of R_m for R134a results in a higher COP. The COP of supercritical and subcritical for R134a at R_w = 1 is 1.75 and 1.35, respectively. Therefore, the supercritical cycle increases the COP by 0.4, an increment of 30%. It can be concluded that R134a is more efficient given that the large value of R_m at $R_w = 1$, and possessing a greater value of COP, 35% greater than that of R245fa.

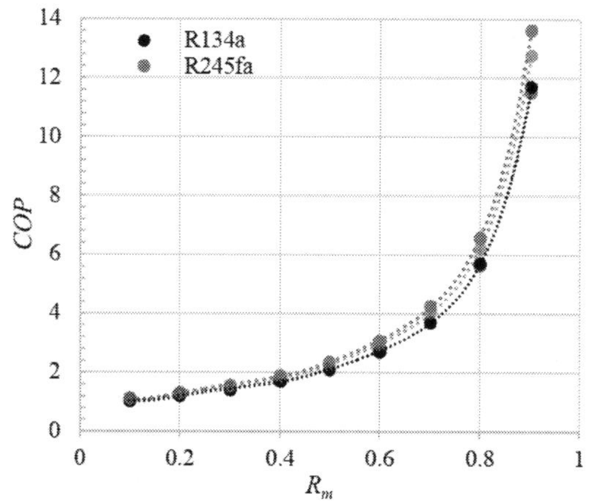

Figure 5. COP as function of ratio R_m for R134a and R245f in subcritical (*faded colour*) or supercritical (*bold colour*) mode.

4 Conclusion

This work presents a mathematical evaluation on the cooling, heating and power performance of a combined system of Organic Rankine Cycle and Vapour Compression Cycle. The work focusses on the effect of changing the mass flow rate split ratio R_m on the energy output and compare the effect of using subcritical and supercritical working fluid. Particular attention is paid toward the scenario where the ratio of the turbine power output to the energy consumption R_w is equal to one, in view of creating a self-powered cooling and heating system. The following conclusions are drawn:

1. R_w decreases exponentially against an increasing R_m at a fixed gradient of approximately $e^{-4.6R_m}$ for both R134a and R245fa. Supercritical cycle is found to be consistently offset the subcritical cycle by a constant magnitude, allowing $R_w = 1$ to be achieved at higher R_m.
2. The net coefficient of performance COP increases exponentially with the increase of R_m for both R134a and R245fa. The COP is higher with the use of R245fa compared to the COP of R134a. The COP is also slightly increased when using supercritical cycle.
3. R134a is recommended as the working fluid for the system, given its larger value of R_m at $R_w = 1$. In supercritical mode, the COP of R134a is 35% higher than that of R145fa.

References

1. T. Kubota, S. Jeong, D.H.C. Toe, D.R. Ossen. Energy consumption and air-conditioning usage in residential buildings of Malaysia. Journal of international Development and Cooperation, **17(3)**, pp.61-69. (2004)
2. T. Wang, Y. Zhang, Z. Peng, G. Shu. A review of researches on thermal exhaust heat recovery with Rankine cycle. Renewable and sustainable energy reviews, **15(6)**, pp.2862-2871. (2011)
3. M. Yari. Exergetic analysis of various types of geothermal power plants. Renewable Energy, **35(1)**, 112-121. (2010)
4. S. Lecompte, H. Huisseune, M. Van Den Broek, B. Vanslambrouck, M. De Paepe, M. Review of organic Rankine cycle (ORC) architectures for waste heat recovery. Renewable and Sustainable Energy Reviews, **47**, pp.448-461. (2015)
5. S. Quoilin, M. Orosz, H. Hemond, & V. Lemort. Performance and design optimization of a low-cost solar organic Rankine cycle for remote power generation. Solar energy, **85(5)**, 955-966. (2011)
6. B.F. Tchanche, G. Papadakis, G. Lambrinos and A. Frangoudakis. Fluid selection for a low-temperature solar organic Rankine cycle. Applied Thermal Engineering, **29(11-12)**, pp.2468-2476. (2009)
7. Y. Dai, J. Wang and L. Gao. Parametric optimization and comparative study of organic Rankine cycle (ORC) for low grade waste heat recovery. Energy Conversion and Management, **50(3)**, pp.576-582. (2009)
8. J. Wang, J., Y. Yang. Energy, exergy and environmental analysis of a hybrid combined cooling heating and power system utilizing biomass and solar energy. Energy Conversion and Management, **124**, pp.566-577. (2016)
9. A.S. Nafey, M.A. Sharaf. Combined solar organic Rankine cycle with reverse osmosis desalination process: Energy, exergy, and cost evaluations. Renewable Energy, **35(11)**, pp.2571-2580. (2010)
10. C. Li, G. Kosmadakis, D. Manolakos, E. Stefanakos, G. Papadakis, & D.Y. Goswami. Performance investigation of concentrating solar collectors coupled with a transcritical organic Rankine cycle for power and seawater desalination co-generation. Desalination, **318**, 107-117. (2013).
11. M. Ebrahimi and I. Moradpoor. Combined solid oxide fuel cell, micro-gas turbine and organic Rankine cycle for power generation (SOFC–MGT–ORC). Energy Conversion and Management, **116**, pp.120-133. (2016)
12. S. Karellas, K. Braimakis. Energy–exergy analysis and economic investigation of a cogeneration and trigeneration ORC–VCC hybrid system utilizing biomass fuel and solar power. Energy conversion and management, **107**, pp.103-113. (2016)
13. T. Kubota, S. Jeong, D.H.C. Toe, & D.R. Ossen. Energy consumption and air-conditioning usage in residential buildings of Malaysia. Journal of international Development and Cooperation, **17(3)**, 61-69. (2011)
14. K. Braimakis, A. Thimo and S. Karellas. Technoeconomic analysis and comparison of a solar-based biomass ORC-VCC system and a PV heat pump for domestic trigeneration. Journal of Energy Engineering, **143(2)**, p.04016048. (2016)

Performance Evaluation of a Resonant-integrated Pumping System

Lee Vincent Chieng-Chen

Faculty of Engineering and Science, Curtin University Malaysia, Sarawak

Abstract. Impedance pump is a simple valve-less pumping mechanism; it offers a low energy, low noise alternative at both macro- and micro-scale devices. It is also demonstrated to be a promising new technique for producing and amplifying net flow. There have been research studying the effects of series-connected impedance pump, where an increase in net flow is exhibited. In this study, an integrated system of conventional pump and impedance pump is introduced. This paper describes the performance evaluation of this integrated pumping system, with emphasis on the amount of amplification induced as a function of Womersley number (normalized excitation frequency) and normalized pressure head. Due to the nature of the resonant valve-less impedance pump, the integrated pumping system exhibits similar behaviour and characteristics as an impedance pump, such as the pulsatile nature of net flow. Results show positive outcomes where maximum amplification of 91.7% is demonstrated at resonance.

1 Introduction

Impedance pump is a very simple design that offers a promising new technique for producing and amplifying a net flow for both macro- and micro-scale devices [1, 2]. The impedance pump is a valve-less pump, which does not require impellers to operate. It is formed by joining a flexible tube to rigid sections. Asymmetrical excitation at a single location of the fluid-filled elastic tube will result in unidirectional flow due to the mismatch in impedance. Such pumping mechanism has shown to be highly sensitive towards the impedance in the tube, the location, and excitation frequency [1, 3-6].

The first demonstration of valve-less pumping through an impedance pump, was demonstrated by Gerhart Liebau in 1954, using an elastic tube connected to reservoirs at different heights [7, 8]. Following Liebau's work in 1954, there have been several research studying the underlying physics of the system, be it numerical or experimental. It was then, that the pump showed possibility in flow amplification. In recent years, study on sequential excitations on a single elastic tube shows promising results where increase in net flow is evident [9]. Other work on flow amplification includes design which incorporates multi-stage impedance pumping system [10] with flow enhancement of 69%.

Numerical study on flow amplification, integrating a steady and oscillating flow is also presented, where an estimated amplification of 120% is demonstrated [11]. The results are, however, not validated with any experimental data. With reference to that study, it shows that amplification of flow rate is indeed achievable with the aid of impedance pump. This study aims to prove this concept experimentally. The focus of this study is the amount of amplification induced, demonstrated as a function of Womersley number (normalized excitation frequency) and normalized pressure head during activation of the impedance pump.

2 Experimental methods and materials

The integrated pumping system utilises a modified version of the conventional impedance pump setup in Ref [10] with flow discharge at variable pressure heads. Figure 1 shows the schematic diagram of a modified two-stage open-loop impedance pump with flow discharge of variable pressure heads. Water inlet is connected in line as the silicone rubber tube to the source reservoir (reservoir on the left) as illustrated in the figure, from a water pump model DB-330A with power rating of 330 Watts with flow rate of 6 L/min. The water pump will transport the water from its source into the reservoir. Multiple flow discharges of variable pressure heads are configured on the deliverable reservoir (reservoir on the right) as illustrated, where the flow rates are measured. The pressure heads are configured at height difference of 0 to 60 mm (pressure head difference of 0 to 735 Pa) with 20 mm interval above the initial water level of 250 mm. These data will serve as the benchmarking data for the integrated pumping system. The elastic silicone rubber tube will then be excited simultaneously while water pump is transporting the water through the discharge. Flow rates, with additional effect from the excitations, are measured and analysed against the benchmarking data for comparison. The analysis of the flow dynamics will be in its non-dimensional form. Detailed analysis method are available in Ref [12].

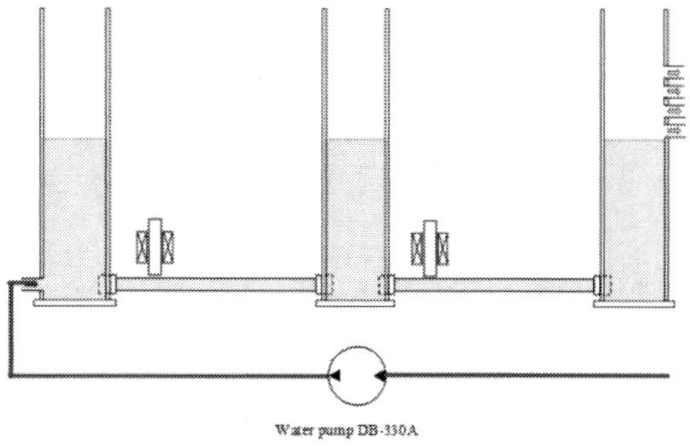

Figure 1. Schematic of experimental setup

3 Results and discussion

Amplifications with respect to different pressure heads, as a function of Womersley number are demonstrated. It is observed that active amplification (i.e. flow higher than 0%) is only within the range of Womersley number 59.4 to 92.0 (2.5 Hz to 6.0 Hz), as shown in Figure 2. This observed range of working amplification is similar to the working range of impedance pump as presented in Ref [10].

The amplified flow is observed to have similar characteristics as impedance driven flow where highly non-linear behaviour is observed. Similar to the impedance driven flow, the effective amplification range is shown to be at the Womersley number 70.3 to 88.1 (3.5 Hz to 5.5 Hz). This suggests that the effective amplification region coincides with the resonance of the system.

This phenomenon is due to the fact that the total flow rate is a summation of water pump flow rate and impedance pump flow rate as expressed in Equation (1)

$$Q_{tot} = Q_{pump} + Q_{impedance} \tag{1}$$

The water pump flow rate observes the Hagen Poiseuille law, as expressed in Equation (2) and impedance pump flow rate is as expressed in Equation (3)

$$Q_{pump} = \frac{\pi d^4}{128\mu l}\Delta P \tag{2}$$

$$Q_{impedance} = \frac{1}{\sqrt{R^2 + S^2}}\Delta P_o \sin(\omega t - \vartheta) \tag{3}$$

where $R = \frac{128\mu l}{\pi d^4}$ $S = \omega L - \frac{1}{\omega C}$

$L = \frac{4\rho l}{\pi d^2}$ $C = \frac{\Delta V}{\Delta(\Delta P_c)}$

R denotes the fluid resistance term, S is the fluid reactance term, L is the fluid inertia term, and C is the tube compliance term. ΔP is the conventional pump pressure difference ΔV is the change in volume and $\Delta(\Delta P_c)$ is the change in pressure change within the elastic rubber tube, ΔP_o is the pressure change due to excitation, ϑ is the phase difference, and l is the length of elastic tubes.

Looking into the individual curve of different pressure heads, the curves show different characteristics as the pressure head changes. The curve of zero pressure head amplification is shown to be similar to the impedance driven flow where peaks are found to be at Womersley number of 59.4 to 102.9. Based on the characteristics shown on the output flow, it is shown that for amplification of flow at zero pressure head, the output flow behaves like impedance driven flow where it has high non-linear dependency towards the excitation frequencies.

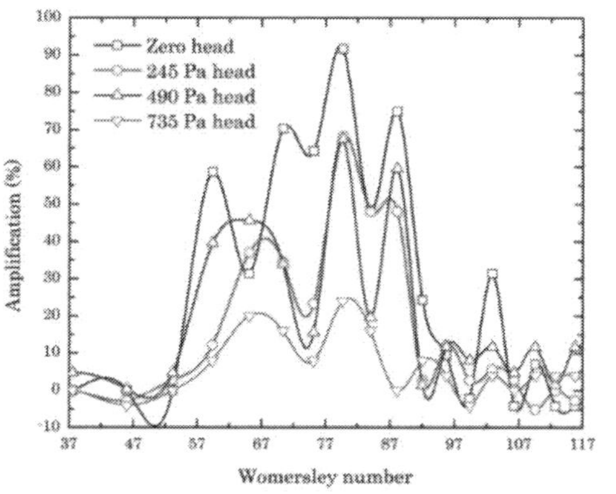

Figure 2. Flow amplification at different pressure heads

For pressure head of 245 Pa, flow is observed to increase as pressure head increases until it reaches the maximum pressure head where flow is observed to decrease from that point onwards. It is, however, observed that the amplified flow, again, peaks at Womersley number of 75.1 to 88.1, which is the resonance region of excitation for impedance pump. Flow is observed to be oscillating with minimal amplification for Womersley number ranges from 92 to 118.9. Amplification is apparent to be lower than the flow induced from zero pressure head. For pressure head of 490 Pa, the flow exhibits better amplification as compared to flow amplified at 245 Pa. Highly non-linear flow is observed at Womersley number of 75.1 to 88.1, maximum flow amplified under the resonant excitation is shown to be similar with the case of 245 Pa. Flow is observed to be oscillating with minimal amplification for Womersley number ranges from 92 to 118.9, similar to the previous case.

Low amplified flow is observed for the pressure head of 735 Pa. The flow is shown to have non-linear behaviour, similar to previous two cases. Amplifications are observed to be extremely low as compared to all previous cases. For all cases, the maximum amplification occurs at Womersley number of 79.7 (4.5 Hz), which is demonstrated to be the resonant excitation frequency of the impedance pump. It is also suggested that the excitation region of Womersley number ≤79.7 is the amplification region; and that excitation region > 79.7 is the suppression region for the system. Hence, a much lower amplification is observed.

With reference to results presented in Figure 2, it is shown that the excitation region of interest is within the Womersley number 75.1, 79.7 and 84. A more in-depth analysis of amplification and pressure head is made on these three points of interest.

Figure 3(a) shows the normalized flow comparison between a normal flow and the amplified flow at Womersley number of 75.1. As shown in the figure, the trend for both curves is observed to be similar. The normalized pressure head characterises the pressure head in the reservoir, and it shows that as the pressure head increases, the flow reduces in an oscillatory manner. The flow is observed to reduce from zero pressure head till 0.1, then slight rise at 0.2 and decrease again at 0.3. Ideally, flow rates of a conventional pump show a linear decrease in magnitude against the pressure head. A sinusoidal behaviour is observed, due to the elastic nature of the tubes, that exhibits a moving boundary as opposed to a rigid tube with fixed boundary. Therefore, it is showed that even with a steady flow from a pump passing through an elastic tube, sinusoidal flow behaviour can be expected. The amplified flow is observed to behave the same way. Margin between the normal and amplified flow is observed to narrow down as the pressure head increases, indicating that the amplification effect reduces as the pressure head increases. Reduction in the amplification effect is seemed to be in a rather linear manner.

Figure 3(b) shows the comparison between the amplified and normal flow at Womersley number of 79.7. Amplification is found to be highest at this Womersley number, where an amplification of 91.7% is demonstrated. Based on the two curves, the margin of amplification is observed to be quite parallel from zero pressure head till normalized pressure head of 0.2. For normalized pressure head of 0.2, the increment gradient of amplified flow is shown to be higher than the normal flow, suggesting a profound amplification effect at that pressure head. For pressure head of 0.3 however, the amplification effect is shown to decrease drastically.

The comparison of amplified and normal flow at Womersley number of 84.0 is shown in Figure 3(c). Amplification effect is observed to be lower than 79.7. The amplification curve is observed to decrease gradually along the normalized pressure head from zero to 0.3; whereas for the normal flow, the trend is observed to reduce from zero pressure head till 0.1, followed by a hike in 0.2 and a drop again at 0.3. The amplification effect is however different from the previous two cases where the margin is shown to be quite parallel from zero pressure head to 0.1. From pressure head of 0.1 to 0.2 however, a drastic narrow margin is shown due to the hike in normal flow but for continual decrement for the amplified flow. From 0.2 to 0.3, the margin is shown to be parallel once again.

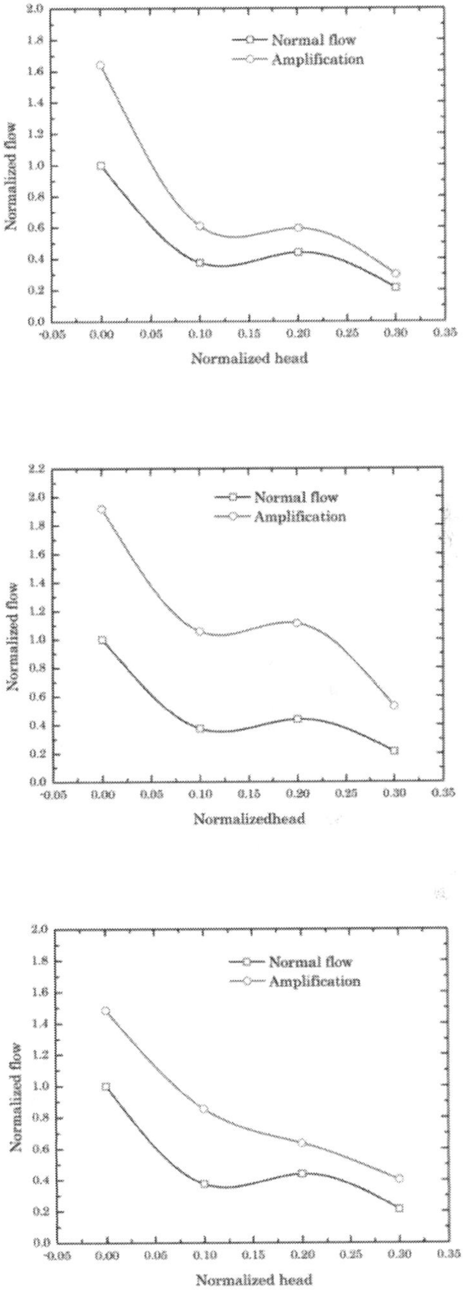

Figure 3. Flow amplification at Womersley number (a) 75.1 (b) 79.1, and (c) 84.0

In comparison to Ref [11] which reported an amplification of approximate 120% at resonance, current experimental study shows that flow amplification through integration of impedance pump is indeed possible. Maximum flow amplification is demonstrated at 91.7% at resonance. This shows that the amplification of flow is rather well estimated. This deviation between the numerical and experimental result could possibly due to the de-coupled modelling between fluid and structure domain in Ref [11].

With reference to Figure 3, the amplified flow rates show some differences in its behaviours. This suggests that maximum amplified flow rate is achievable when the excitation frequencies coincide with the elastic tubes' natural frequency. This is clearly shown in Figure 3(a) and (b), where the amplified flow behaviours are similar to the normal flow. Hence, the excitation frequency plays an important role in determining the system's efficiency.

With reference to Equation (1) – (3), it is shown that the total flow rate is highly dependent on the impedance pump. Flow rate resulted from the water pump is basically constant throughout the whole process of the same pressure head. Flow rate from the impedance pump, is however, change with respect to time. Hence, a pulsation of flow will occur. It is shown, as well, that the amplification percentage reduces as the pressure head increases.

4 Conclusion

This study presents and introduces an integrated system of conventional pumping system and impedance pump for efficient pumping. Profound flow amplification is observed as an effect of the integration at the working frequencies region of 3 Hz to 6 Hz or Womersley number of 75.1 to 92.0. Impedance driven flow characteristics are observed in the amplified flow such as the sensitivity of flow towards the excitation frequencies and the non-linear behaviour of the flow. Maximum amplification is observed to be at resonant frequency of 4.5 Hz with an amplified flow of 91.7%. It is also suggested that the optimum pressure head for effective amplification in this study is 490 Pa.

References

1. I. Avrahami and M. Gharib, Computational studies of resonance wave pumping in compliant tubes, J Fluid Mech 608 (2008) 139-160.
2. D. Rinderknecht, A. I. Hickerson and M. Gharib, A valveless micro impedance pump driven by electromagnetic actuation, J Micromech Microeng 15 (2005) 861-866.
3. V. C.-C. Lee, Y. A. Abakr and K.-C. Woo, Dynamics of Fluid in Oscillatory Flow: The Z Component, Journal of Engineering Science and Technology 10(10) (2015) 1361-1371.
4. C. G. Manopoulus, D. S. Mathioulakis and S. G. Tsangaris, One dimensional model of valveless pumping in a closed loop and a numerical simulation, Phys Fluids 18 (2006) 017106.
5. A. I. Hickerson, An experimental analysis of the characteristic behaviours of an impedance pump, Ph.D. Thesis, California Institute of Technology, 2005.
6. S. Timmermann and J. T. Ottesen, Novel characteristics of valveless pumping, Phys Fluids 21 (2009) 053601.
7. G. Liebau, Über ein ventilloses pumpprinzip. Naturwissenschaften 41(1954) 327. (in German)
8. G. Liebau, Die stromungsprinzipien des herzens. Zietschrift f'r Kreislauf-forschung 44 (1955) 677–84. (in German)
9. M. Rosenfeld and I. Avrahami, Net flow rate generation by a multi-pincher impedance pump. Comput Fluids 39(9) (2010) 1634-1643.
10. V. C.-C. Lee, H. S. Gan, Y. A. Abakr and K.-C. Woo, Bulk Flow Behaviour of a Two-Stage Impedance Pump. Engineering Letters 22 (2014) 53-62.
11. V. C.-C. Lee, Y. A. Abakr and K.-C. Woo, Impedance Pumping System as Flow Amplification Device, Oral Presentation. In: 2011 International Conference on Applied Mechanics, Materials and Manufacturing, China, 2011.

A novel active disturbance rejection controller for an electromagnetic valve actuator

Jiayu Lu and Siqin Chang

School of Mechanical Engineering, Nanjing University of Science and Technology, Nanjing, China

Abstract. In order to achieve fully variable and precise motion control of the electromagnetic valve actuator, enhance the engine performance, a novel active disturbance rejection controller for an electromagnetic valve actuator based on trajectory planning and acceleration feedforward is proposed. A fourth-order trajectory planning is used to achieve fully variable valve motion control, including variable valve lift and timing. It can also reduce the impact and vibration of EMVA system. Active disturbance rejection controller is used to estimate the variant dynamic and external disturbances of the system. The acceleration feedforward is compensated for the system to improve the tracking and steady state accuracy. Comparative simulations results show the proposed controller can improve the dynamic performance and the robustness of the system, and enhance the control precision.

1 Introduction

Camless engine is considered to be an efficient technology to improve the engine fuel economy, emissions and torque output performance. A novel electromagnetic valve actuator (EMVA) developed by the team of author's which has the characteristics of high response and large driving force is used on the camless engine. The valve lift, the valve duration, and the valve timing can be regulated flexibly by controlling the motions of the EMVA. And it can greatly improve the performance of the engine [1]. However the valve is directly driven by the EMVA, there is no intermediate link buffer, various disturbance factors are directly reflected on the system, which makes the dynamic characteristics of the system complex and changeable.

The main control goal of EMVA system is to achieve fully variable and precise motion control in spite of the various disturbance. Furthermore, soft landing control is required for the controller in order to guarantee working life of the gas regulating valve and reduce the working noise [2-5].

Active disturbance rejection controller (ADRC) is a new control method proposed by Han [6]. In ADRC, various known and unknown factors that influence the system performance are lumped together as total disturbance, and they are estimated and compensated in real time via an extended state observer (ESO), which is a key component of ADRC [7].

In this paper, a modified active disturbance rejection controller (MADRC) is proposed to meet the control requirements of EMVA system, including high control accuracy, robustness and anti-disturbance ability. A fourth-order trajectory planning is used to achieve fully variable valve motion control, including variable valve lift and timing. It can also reduce the impact and vibration of EMVA system. Active disturbance rejection controller is used to estimate the variant dynamic and external disturbances of the system. The acceleration feedforward is compensated for the system to improve the tracking and steady state accuracy. Simulation results indicate the effectiveness of the proposed controller.

2 Structure and mathematical model of EMVA

The structure of EMVA is shown in Fig. 1. The EMVA consists of a moving coil, permanent magnets, a valve, an inner core, and an outer core. The moving coil of EMVA is connected to the valve. The inner and outer core form a uniform air gap magnetic field. The movement of the valve can be controlled by controlling the magnitude and direction of the coil current [8].

The EMVA consists of electrical, magnetic, and mechanical subsystems, which can be described as follow

$$\begin{cases} \dot{x} = v \\ \dot{v} = (K_m/m)\cdot i - (K/m)\cdot x - F_f/m - F_d/m \\ \dot{i} = -(R/L)\cdot i - (K_e/L)\cdot v + u/L \end{cases} \quad (1)$$

where u is the input phase voltage. i is the current through the coil. R , L represent the resistance and inductance of the coil respectively. m is the moving mass, x denotes the mover position, v is the velocity of the mover. K_m is the force sensitivity if the actuator, which denotes the ratio of the electromagnetic force to the input current. K_e is the back electromotive force coefficient.

F_f, F_d represent the friction and the external disturbance respectively, which is the uncertain nonlinearities of the system. Furthermore, in this EMVA, K_m and K_e are approximately equal in value.

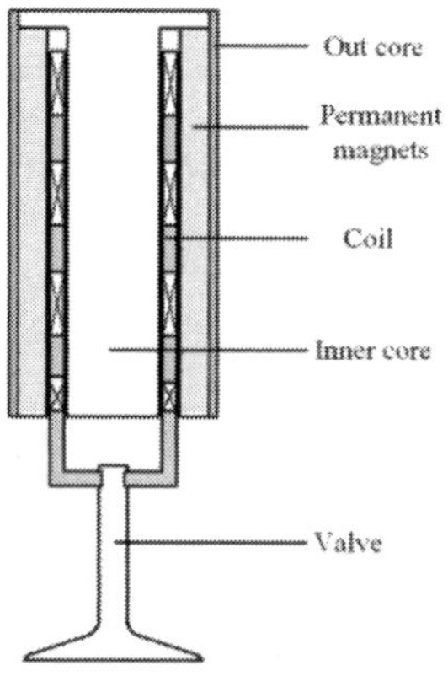

Figure 1. Structure of EMVA

3 Control System Design of EMVA

3.1 Basic ADRC

The basic ADRC consists of three parts: Tracking-Differentiator (TD), Extended State Observe (ESO) and Nonlinear Proportional-Derivative controller (NPD). Consider the following nonlinear second-order plant:

$$\begin{cases} \dot{x}_1 = x_2 \\ \dot{x}_2 = f(x_1, x_2, d(t), t) + b \cdot u \\ y = x_1 \end{cases} \tag{2}$$

where y is the output to control, u is the control input, b is the system parameter, and $f(x_1, x_2, d(t), t)$ is the nonlinear total disturbance. The control objective is to make the output y tracks the desired signal y_d as far as possible in spite of the nonlinear disturbance.

The TD is used to arrange the transitional trajectory of the desired signal y_d.

ESO is the key component of ADRC, which can estimate the total disturbance. The discrete-time form of ESO for second-order plant Eq. (2) with sampling period h is

$$\begin{cases} e = z_1(k) - x(k) \\ z_1(k+1) = z_1(k) + h \cdot (z_2(k) - \beta_{01} \cdot e) \\ z_2(k+1) = z_2(k) + h \cdot (z_3 - \\ \qquad \beta_{02} \cdot fal(e, \alpha_1, \delta) + b \cdot i_d(k)) \\ z_3(k+1) = z_3(k) - h \cdot \beta_{03} \cdot fal(e, \alpha_2, \delta) \end{cases} \tag{3}$$

where i_d is the control input of the second-order system, β_{01}, β_{02} and β_{03} are observer gains, z_1、 z_2 ,and z_3 are estimates of the output x, the derivative of x, and the total disturbance respectively.

The nonlinear function $fal(e, \alpha, \delta)$ is defined as

$$fal = \begin{cases} e \cdot \delta^{\alpha - 1}, & |e| \le \delta \\ |e|^{\alpha} \cdot \mathrm{sgn}(e), & |e| > \delta \end{cases} \tag{4}$$

Parameter α satisfy condition $\alpha < 1$, δ satisfy condition $\delta = j \cdot h$, where j is a positive integer.

The indirect control input u_0 is determined by NPD which is given as

$$\begin{cases} e_1 = y_1 - z_1 \\ e_2 = y_2 - z_2 \\ u_0 = \beta_1 fal(e_1, \alpha_1, \delta) + \beta_2 fal(e_2, \alpha_2, \delta) \end{cases} \tag{5}$$

where β_1, β_2, α_1, α_2 and δ are controller parameters. Finally, the control input u is given as

$$u = \frac{u_0 - z_3}{b} \; . \tag{6}$$

3.2 Proposed modified ADRC

The transitional trajectory of the desired signal *yd* arranged by TD is shown in Fig. 2 which is a kind of second-order profile planning.

As Fig. 2 shows, second-order profile planning can achieve good fastness of the control. However, the accelerated speed in the second-order profile planning is mutated which may cause the system to vibrate or oscillate, and it is not conducive to the precision control performance requirement. In order to solve this problem a forth-order profile planning is proposed. The forth-order profile planning can achieve smooth acceleration and deceleration, which can avoid mechanical shock and oscillate effectively. The typical forth-order profile planning for the point-to-point motion is shown in Fig. 3. T_0, s represent the movement time and lift of the EMVA respectively. *dmax* is the max jerk rate, td is the max jerk rate period. t_a is the max acceleration period, and *tj* is the max jerk period.

The use of sections to divide the text of the paper is optional and left as a decision for the author. Where the author wishes to divide the paper into sections the formatting shown in Table 2 should be used.

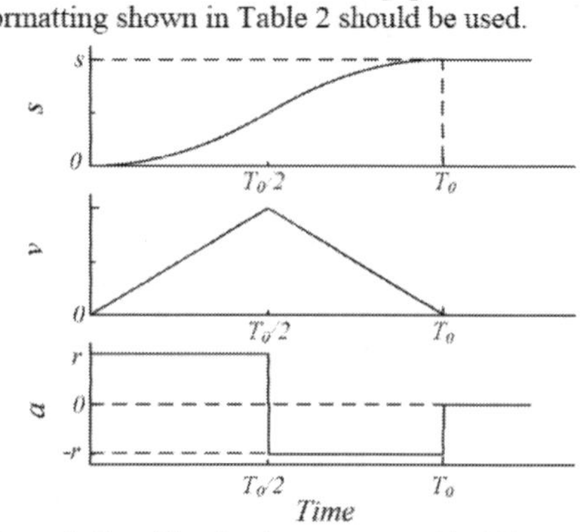

Figure 2. Transitional trajectory arranged by TD

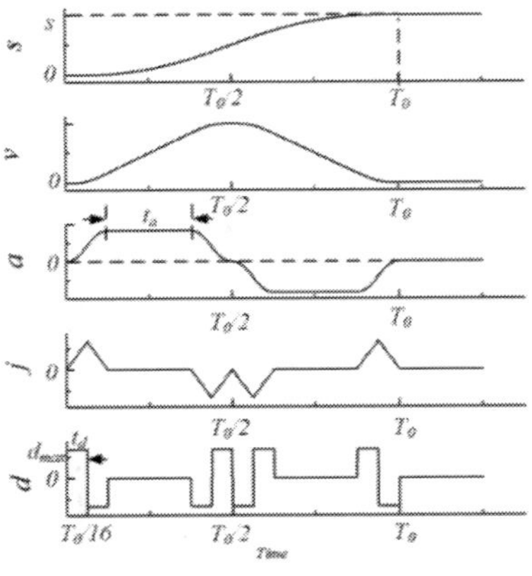

Figure 3. Typical forth-order profile planning

As Eq. (1) shows that the accelerated speed is mainly determined by the coil current, which is limited by the working voltage and actuator performance. Meanwhile, the max velocity of the actuator is not limited, according to [9], t_j is tend to be zero. The movement distance s can be given as

$$s = (t_d^{\,2} \cdot t_a^{\,2} + 6t_d^{\,3} \cdot t_a + 8t_d^{\,4}) \cdot d_{\max}. \tag{7}$$

In order to achieve the full flexibility control of the EMVA, T_0, s should be variable. Let $t_d = T_0/16$ and $t_a = T_0/4$, so that $d_{\max} = 4096 \cdot s/(3 \cdot T_0^{\,4})$, d_{max} is only determined by T_0 and s. The date of the trajectory planning is calculated by digital integration which can be described as

$$\begin{cases} j(t) = j_0 + d_{\max} \cdot t \\ a(t) = a_0 + j_0 \cdot t + d_{\max} \cdot t^2/2 \\ v(t) = v_0 + a_0 \cdot t +_0 j \cdot t^2/2 + d_{\max} \cdot t^3/6 \\ s(t) = s_0 + v_0 \cdot t + a_0 \cdot t^2/2 + j_0 \cdot t^3/6 + d_{\max} \cdot t^4/24 \end{cases} \tag{8}$$

The feedback control is based on the system deviation which is caused by the output changes. So there exists time-delay in the feedback control. For the high speed and high precision control, the time delay of desired trajectory tracking may be more obvious and a steady-state error may be exist, if the simple feedback control is used. Meanwhile, the feed-forward control can solve this problem. In order to improve the control performance of the system, a compound controller with feed-forward and feedback control is presented.

Fig. 4 shows the structure diagram of feedback controller, where $G_c(s)$, $G_p(s)$ represent the controller transfer function and controlled plant transfer function respectively, w is the total disturbance of the system, z_3 is the total disturbance which is estimated by the ESO. The system transfer function can be described as

$$G_1(s) = \frac{G_c(s) \cdot G_p(s)}{1 + G_c(s) \cdot G_p(s)} \tag{9}$$

and the error transfer function of the system is

$$E_1(s) = \frac{1}{1 + G_c(s) \cdot G_p(s)} \tag{10}$$

which is inevitable.

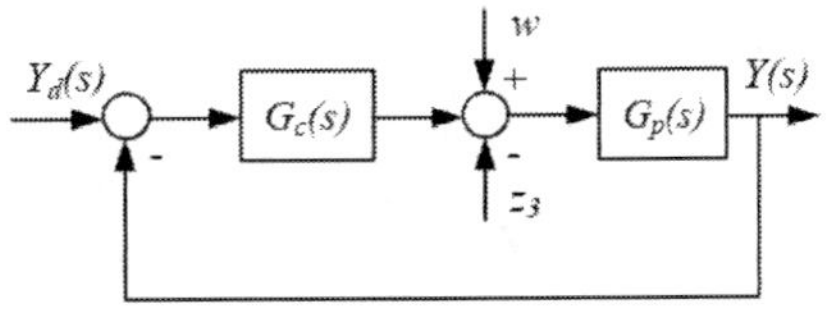

Figure 4. Structure diagram of feedback controller

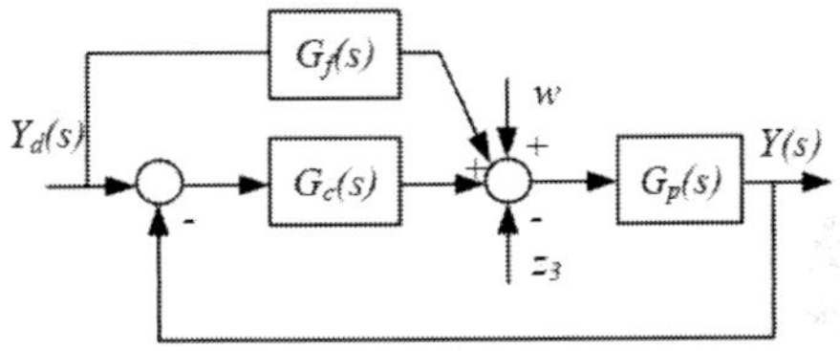

Figure 5. Structure diagram of compound controller

The structure diagram of compound controller with feed-forward and feedback control is shown in Fig. 5. The system transfer function can be described as $G_2(s) = \dfrac{G_c(s) \cdot G_p(s) + G_f(s) \cdot G_p(s)}{1 + G_c(s) \cdot G_p(s)}$, where $G_f(s)$ is the feed-forward controller transfer function. And the error transfer function of the system is $E_2(s) = \dfrac{1 - G_f(s) \cdot G_p(s)}{1 + G_c(s) \cdot G_p(s)}$.

According to Eq. (6), if z_3 can completely estimate w, then $G_p(s) = \dfrac{1}{s^2}$. The feed-forward control transfer function is s^2, if acceleration feed-forward is used on the control system. So the error transfer function of the system tend to be zero, which proves that the compound controller can improve the control precision and does not affect the stability of the system.

The proposed control system of EMVA based on a novel ADRC consists of current and position loops, which is shown in Fig. 6. The modified ADRC is used in the position loop, the basic ADRC is implemented in the current loop.

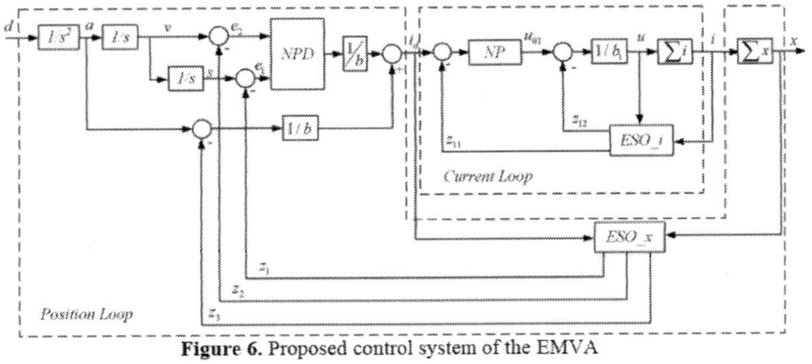

Figure 6. Proposed control system of the EMVA

 https://doi.org/10.1051/matecconf/201820202010

4 Simulation results

To test the effectiveness of the proposed modified ADRC, comparative simulations of the novel ADRC and traditional ADRC are implemented on EMVA with parameters given in Table I using MATLAB/Simulink.

4.1. Comparative simulations

he comparative simulations results are shown in Fig. 7. As Fig. 7(a) shows that both controllers have high steady precision, and there is no overshoot happened, but the MADRC achieves better performance for fast response. Fig. 7(b) is the tracking errors of the system, as we can observe in Fig. 7(b) that the max tracking error of ADRC and MADC is 159 μm and 9 μm respectively, which illustrates that the MADRC achieves better tracking performance. Fig. (c) is the velocity of EMVA.

It is demonstrated that a seating velocity (the average of three sampled valve velocities before the valve reaches the closed position) of 0.005 m/s has been achieved in the MADRC system. And the average seating velocity for ADRC system is about 0.02 m/s, which illustrates that the MADRC achieves better soft landing performance. The control current of the system is shown in Fig. 7(d). As we can see that the control current of MADRC system is smoother than the ADRC system's, which may make the system more stable.

4.2 Simulation results

In order to achieve the fully variable control of the EMVA, the movement time T0, and lift s should be variable.

Fig. 8 (a) shows the system response with respect to various reference valve lift. The reference valve lifts are set to 4 mm, 6 mm, and 8 mm respectively. The movement profile are planned by the fourth-order profile planning which is introduced in section 3.2.1. The movement time are set to 8 ms. Fig. 8(b) shows the system response with respect to various movement time at the lift of 8 mm. The movement times are set to 6 ms, 8 ms, and 10 ms respectively. Fig. 8 demonstrate that the MADRC achieves better performance for control precision and fast response, and the position error tends to be zero when MADRC is used in EMVA system.

Table 1. Parameters of the EMVA

Parameter	Valve
R	1.38Ω
L	$0.9\,mH$
K_e	$12\,V\bullet s/m$
K_m	$12\,N/A$
m	$94.6\,g$
u	$30\,V$

4.3 Anti-disturbance ability analysis

The EMVA suffers from various disturbances, such as combustion pressure and supercharge pressure. Since the disturbances are not well defined in the investigation, a step disturbance F is applied on the EMVA for a time 3-5 ms to estimate the disturbance rejection of the systems. Fig. 9 shows the simulation results when the values of F is ±100 N. There is no overshoot happen in both ADRC and MADRC system. Fig. 9 (b)-(c) show the error of the system. Here error= $x-\bar{x}$, where x represents the trajectory obtained with control without disturbance, represents the trajectory obtained with control with disturbance. As Fig. 9 (b)-(c) shows that the MADRC has a smaller error than the basic ADRC, which shows that the MADRC has a better anti-disturbance ability than the basic ADRC.

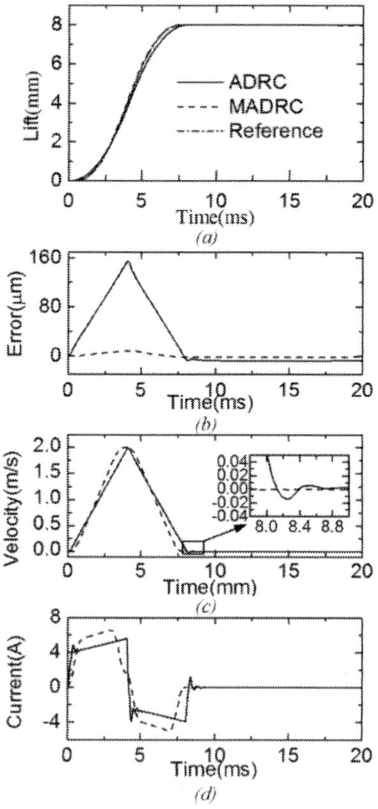

Figure 7. Comparative simulations results

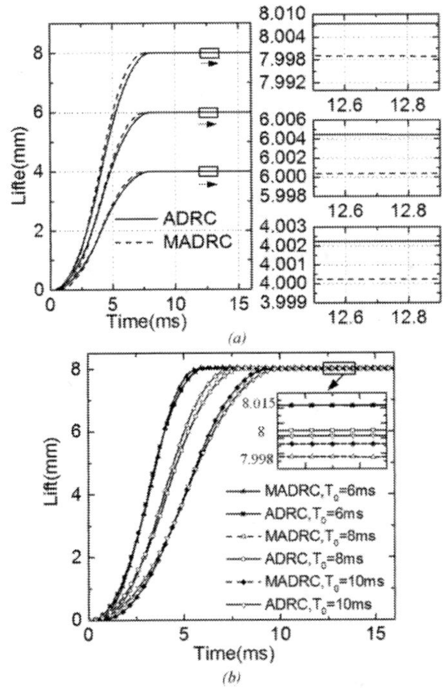

Figure 8. System response with various trajectory

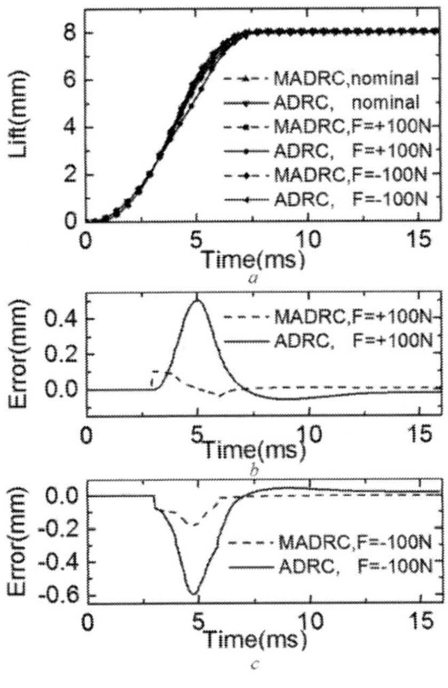

Figure 9. Disturbance rejection

5 Conclusions

A modified ADRC based on fourth-order trajectory planning and acceleration feedforward has been developed in this paper to achieve high-performance precision valve motion control. The proposed novel ADRC can achieve fully variable, including variable valve lift and timing, and soft landing, precise valve motion control and soft landing control. Meanwhile the proposed novel ADRC has strong robustness and anti-disturbance ability. The performance of the proposed modified ADRC is verified by the comparative simulations. Consequently, if EMVA system with the proposed modified ADRC can be used on an automotive engine, it is predictably that the engine performance will be enhanced.

References

1. Chang, S., and L. Liu. A moving coil permanent magnet linear actuator with high power density, China patent, CN101127474B(2010)
2. Y Yang, Y. P., Liu, J. J., Ye, D. H., Chen, Y. R., & Lu, P. H.. Multi-objective optimal design and soft landing control of an electromagnetic valve actuator for a camless engine. *IEEE/ASME Transactions on Mechatronics*, vol. 18, pp. 963-972(2013)
3. Mercorelli, P. Robust adaptive soft landing control of an electromagnetic valve actuator for camless engines. *Asian Journal of Control*, vol. 18, pp. 1299-1312 (2013)
4. Seethaler, R., Koch, C. R., Chladny, R., & Mashkournia, M. Closed loop electromagnetic valve actuation motion control on a single cylinder engine. SAE Technical Papers, 2 (2013)
5. Paden, B. A., Snyder, S. T., Paden, B. E., & Ricci, M. R. Modeling and control of an electromagnetic variable valve actuation system. IEEE/ASME Transactions on Mechatronics, 20(6), 2654-2665(2015)
6. Han, Jingqing, From PID to active disturbance rejection control, *IEEE transactions on Industrial Electronics*, vol. 56, pp. 900-906.
7. Shi, X., & Chang, S.. Precision motion control of a novel electromagnetic linear actuator based on a modified active disturbance rejection controller. *Proceedings of the Institution of Mechanical Engineers Part I Journal of Systems & Control Engineering*, vol. 226, pp. 606-614(2011)
8. Liu, L., & Chang, S. Motion control of an electromagnetic valve actuator based on the inverse system method. *Proceedings of the Institution of Mechanical Engineers Part D Journal of Automobile Engineering*, vol. 226, pp. 85-93(2011)
9. Yang, Z. Research on fourth-order profile planning algorithm for high accuracy point-to-point motion system. *China Mechanical Engineering*, vol. 18, pp. 2346-2339(2007)

https://doi.org/10.1051/matecconf/201820202011

Individual Cylinder Air-Fuel Ratio Estimation for a Gasoline Engine with Electromagnetic Valve Train

Yaxuan Xu and Siqin Chang

Department of Mechanical Engineering, Nanjing University of Science and Technology, Nanjing, China

Abstract. For the multi cylinder gasoline engine, the consistency of each cylinder is an important index to affect the emission and the power. In this paper, in order to reduce the air-fuel ratio (A/F) maldistribution of the engine based on the electromagnetic valve train (EMVT), an individual cylinder A/F estimation algorithm is proposed for the individual cylinder A/F control. Based on the analysis of the hybrid and transfer models of the exhaust of each cylinder in steady state, an individual A/F observer is established by using Kalman filter algorithm. Then the unknown parameters in the observer are identified by the differential evolution(DE) algorithm. Only a single wide area exhaust oxygen(UEGO) sensor is needed to identify the unknown parameters and estimate the A/F of each cylinder. The combined simulation of GT-Power and Simulink validates the effectiveness of the proposed estimation approach. The results show that the proposed method can provide good estimation results under steady-state condition.

1 Introduction

A/F is one of the most important control variables in the electronic control system of engine, and its control precision is closely related to engine power, economy and emission [1,2]. However, traditional A/F control methods can only regulate the A/F at the confluence point. The A/F imbalance of different cylinders in a multi-cylinder engine is unavoidable. There are many reasons for the A/F maldistribution of each cylinder, and the most direct and main sources of A/F deviations are the cylinder air charges maldistribution and the difference between cylinder injectors [3].

In order to eliminate the difference of A/F between cylinders, the researchers put forward different control methods. Accurate measurement or estimation of individual A/F is the key to the whole A/F control strategy. A common method is to estimate the A/F of each cylinder according to a single UEGO sensor signal placed at the confluence point of the exhaust manifold. According to the different treatment methods on the signal, there are two kinds of technical routes: Time domain method and frequency domain method. The time domain method is to establish the gas mixing process of each cylinder and the dynamic model of the sensor, and regulate the individual A/F calculating from the mixture A/F. The frequency domain method is to extract the frequency components that coincide with the exhaust frequency in the UEGO sensor signal, and to eliminate the difference between cylinders' A/F by suppression of these spectral components [4-6]. Currently both time-domain modeling and frequency-domain modeling require installing an UEGO sensor at the exhaust outlet of each cylinder in the process of parameter identification, and have high consistency requirements for the response speed and delay characteristics of each sensor, increasing the cost and complexity of the test.

This paper focuses on the individual cylinder A/F estimation based on the EMVT. In view of the imbalance of A/F in four-cylinder gasoline engine with the EMVT, a method of estimating the individual cylinder A/F in steady state condition is presented. In this paper, a time domain model is established to generate and transfer the A/F of each cylinder, and a method of parameter identification using DE is proposed [7]. The identification of the model can be accomplished by using only one UEGO sensor at the manifold junction points to simplify the test process and reduce the cost. The Kalman Filter Observer is constructed by using the obtained parameters. The effectiveness of the identification and estimation methods is verified by the combination simulations of GT-Power and Simulink.

2 System model

2.1 System Analysis

The EMVT engine is modified on the basis of a 1.8L four-cylinder port-injected gasoline engine with symmetrical exhaust manifolds, and a UEGO sensor is installed at the confluence of the exhaust manifold. The specification of the targeted engine is shown in Table I. Cylinder head of the engine was reprocessed to fit the installation of EMVT, as shown in Fig. 1. As a fully flexible engine valve train, the EMVT can adjust the air quality into each cylinder by adjusting the opening duration of the intake valve. In the EMVT developed by the task group, each valve is driven separately by the electromagnetic actuator, which can realize the full flexibility adjustment of valve lift, valve timing and valve movement [8-10]. The displacement curve of the performance test experiment of the electromagnetic actuated valve is shown in Fig. 2.

Table 1. Specification of the Engine

Parameter	Value
Displacement	1.8L
Bore	80mm
Stoke	89mm
No. of cylinders	4
Compression ratio	10.5
Rated power	95.7kW@6000rpm
Rated torque	171.3Nm@4500rpm

Figure 1. Test bench of engine cylinder head with EMVT

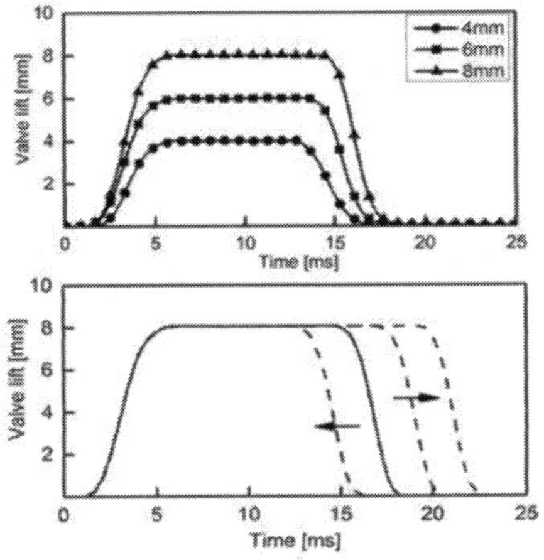

Figure 2. Test curve of valve lift and opening duration of EMVT.

In order to estimate the A/F of each cylinder, an EMVT engine model, a mixture model of exhaust gas, a UEGO sensor model and uneven generation and transfer models need to be constructed. GT-power is used to build the one dimensional model of the prototype engine. Under steady state conditions, the oil film deposition and evaporation can be balanced, and the amount of oil in the cylinder is equal to the amount of injection so that the oil-film model may be omitted in the steady-state control of the cylinder A/F.

2.2 Exhaust Gas-Mixing Dynamics Model

The gas mixture at the exhaust manifold confluence point at a certain time is formed by the exhaust gas from each cylinder. It can be introduced that the exhaust gas flow through the exhaust manifold changes with a simple periodic function under the constant speed condition. Our research object is a four-cylinder engine with a symmetrical exhaust manifold. Under the same working conditions, the proportional coefficients of the fixed rotation angles of each cycle are the same. We select the exhaust bottom dead centre (BDC) of each cylinder as the sampling point. The A/F at the confluence point of exhaust manifold at instant k can be expressed as the following discrete equations:

$$L_{mix}(k)=\sum_{i=1}^{4}c_i\,L_{cyl}(k-i+1)=C\begin{bmatrix}L_{cyl}(k)\\ L_{cyl}(k-1)\\ L_{cyl}(k-2)\\ L_{cyl}(k-3)\end{bmatrix} \quad (1)$$

$$C=\begin{bmatrix}c_1 & c_2 & c_3 & c_4\end{bmatrix} \quad (2)$$

Where $L_{mix}(k)$ represents the A/F of the gas at the confluence point of the exhaust manifold at instant *k*, and $L_{cyl}(k)$ represents the A/F of the gas at exhaust port of current exhausting cylinder. For the four-cylinder engine we have studied with a symmetrical exhaust manifold, the vector ***C*** corresponding to each exhaust BDC is basically the same.

2.3 UEGO sensor model

There is a transmission delay in the process of actual exhaust transmission to UEGO sensor installation point, and the sensor has dynamic effect. The UEGO sensor can be modeled as a first-order system with a propagation delay T_{lag} and a time constant τ as shown in (3).

$$G(s)=\frac{1}{T_{lag}s+1}e^{-\tau s} \quad (3)$$

2.4 Uneven generation and transfer model

In the case of identification parameters using a single sensor, the data available is only the initial intake valve closing time of each cylinder and the final A/F of the UEGO sampling output. Therefore, in order to obtain the proportional coefficient matrix C, it is necessary to analyze the whole process from the specified valve closing time to the A/F of the mixing gas.

In the camless engine without throttle, the pressure in the intake manifold is very close to the atmospheric pressure, and the change rate of the gas pressure in the manifold is very small. Thus, the gas flow through the throttle valve is approximately the same as the gas flow through the intake port. From the engine charge efficiency and the ideal gas state equation, the air mass flow in the intake port of four-cylinder four-stroke gasoline engine can be obtained by

$$\dot{m}_{air} = \frac{V_s \eta_v p_i n}{120 r T_i} \tag{4}$$

Where Vs, ηv, r, Ti, n are engine displacement, engine charge efficiency, gas constant, intake air temperature, and engine speed respectively. By integrating it in the intake stroke of the engine, the intake volume of each cylinder per cycle can be expressed as

$$m_{ac} = \int_{i_o}^{i_c} \dot{m}_{air} dt \tag{5}$$

Where i_o, i_c are intake valve opening time and intake valve closing time respectively. Ideally, the intake air quality of each cylinder per cycle can be expressed as

$$m_{ac} = \frac{\dot{m}_{air}\left[\frac{kg}{h}\right]\left[\frac{h}{3600s}\right]}{n\left[\frac{rev}{\min}\right]\left[\frac{\min}{60s}\right]\left[(n_{cyl}/2)stroke\cdot rev^{-1}\right]} = \frac{n_{cyl}}{30n}\dot{m}_{air} \tag{6}$$

Where n_{cyl} is the number of cylinders. However, due to differences in length and structure of the intake manifold, the actual intake air quality per cycle of each cylinder is not exactly the same. In order to accurately estimate the A/F of each cylinder, the intake distribution deviation should be taken into account. The effect of deviation can be modelled as the cylinder effective coefficient K_i,

$$m_{ac,i} = \overline{K}(1+K_i)\dot{m}_{air}(d_{ur}, n), \quad i=1, 2, 3, 4 \tag{7}$$

Where $\overline{K}$ is a constant coefficient, and K_i is used to obtain the real air charge per stroke going into the i-th cylinder from the air mass flow exiting the intake manifold. The $\dot{m}_{air,n}$ is related to the engine speed and intake valve opening duration. As shown in Fig. 3, the intake air mass flow of the engine under different intake valve opening durations is measured at the speed of 1600 r/min.

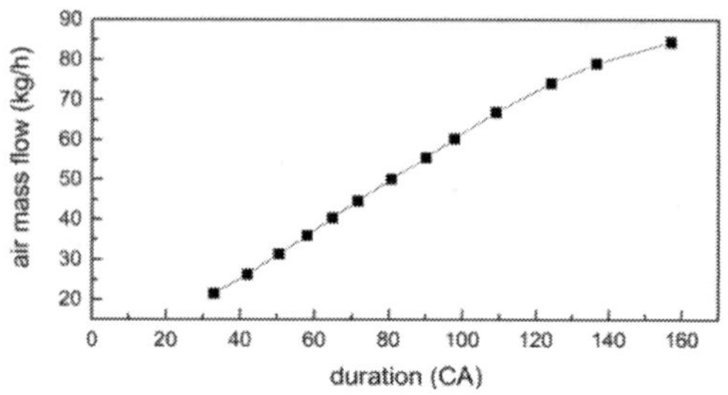

Figure 3. Air mass flow of different intake valve opening durations.

The fitting of cubic polynomial functions with the least squares method is used to fit the curve, and the fitting result is

$$\dot{m}_{air,n=1600} = (-0.0000209)d_{ur}^{\ 3} + 0.00444 d_{ur}^{\ 2} + 0.31 d_{ur} + 7.0848 \tag{8}$$

Where $\dot{m}_{air,n}$ is the intake air flow measured in the intake manifold at a certain speed n and d_{ur} is the intake valve opening duration, where $d_{ur}=i_c-i_o$. The result of curve fitting is shown in Fig. 4. The result shows that the maximum relative error is less than 4%.

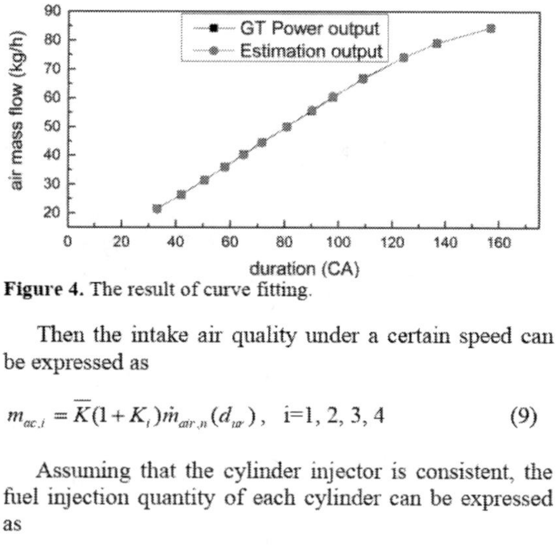

Figure 4. The result of curve fitting.

Then the intake air quality under a certain speed can be expressed as

$$m_{ac,i} = \overline{K}(1+K_i)\dot{m}_{air,n}(d_{ur}), \quad i=1, 2, 3, 4 \tag{9}$$

Assuming that the cylinder injector is consistent, the fuel injection quantity of each cylinder can be expressed as

$$m_f = G_{fs} t \tag{10}$$

Where G_{fs} is the average flow rate of each cylinder injector, and t is the injection pulse width. Under steady conditions, the oil film effect is not taken into account, and the actual A/F of each cylinder is

$$L_{cyl,i} = m_{ac,i} / m_f \tag{11}$$

Thus, the A/F at the gas junction can be calculated according to the gas mixing time domain model described above. Finally, considering the characteristics of the UEGO sensor, the recursive form between the measurement value of the A/F and the real A/F can be obtained by discretization of the Tustin transform as below.

$$L_{sen}(k) = \frac{2T_{lag} - T_s}{2T_{lag} + T_s} L_{sen}(k-1) + \frac{T_s}{2T_{lag} + T_s}\left[L_{mix}(k-d) - L_{mix}(k-d-1)\right] \quad (12)$$

Where $L_{sen}(k)$ represents the A/F value measured by UEGO sensor in the manifold at instant *k*, T_s is the sampling period, and d is the number of delayed cycles, $d=\tau/T_s$. As a result, the modelling of the transfer process from the intake valve opening duration of each cylinder to the measurement value of A/F at the confluence point is completed.

3 Individual cylinder A/F estimation and parameter identification

3.1 Establishment of Kalman filter Observer

Kalman filter algorithm is a continuous "predictive-update" recursive calculation process, and it is efficient in real-time operation. The following is the basic equations of the discrete Kalman filter, in which the first two are the time-updating equations, and the latter three are the state-updating equations [11].

$$\begin{cases} \hat{x}_k^- = A\hat{x}_{k-1}^- \\ P_k^- = AP_{k-1}A^t + Q \\ K_k = P_k^- C^T(CP_k^- C^T + R)^{-1} \\ \hat{x}_k = \hat{x}_k^- + K_k(Y_k - C\hat{x}_k^-) \\ P_k = (I - K_kC)P_k^- \end{cases} \quad (13)$$

Where, Pk is the estimation error; Kk is the kalman filter gain; Q and R are covariance matrices of system noise and measurement noise respectively. The Kalman filter estimates the current system state based on the previously calculated value and the current observational data. We take the A/F of exhaust events' cylinders in the last cycle as the system state X:

$$X(k) = \left[L_{cyl}(k), L_{cyl}(k-1), L_{cyl}(k-2), L_{cyl}(k-3)\right]^T \quad (14)$$

And the mixing A/F measured by the UEGO sensor at the confluence point of the exhaust manifold is taken as the system observation Y:

$$Y(k) = L_{sen}(k) \quad (15)$$

The system state space model is established as follows:

$$\begin{cases} X(k+1) = AX(k) + W \\ Y(k+1) = CX(k+1) + V \end{cases}, \quad A = \begin{bmatrix} 0 & 0 & 0 & 1 \\ 1 & 0 & 0 & 0 \\ 0 & 1 & 0 & 0 \\ 0 & 0 & 1 & 0 \end{bmatrix} \quad (16)$$

Where, A is the state transition matrix of state space; C is the weighting factor matrix, W and V are system noise and measurement noise respectively.

3.2 Parameter identification

From the above model, we can see that in order to calculate the A/F of each cylinder, we should find out the parameters of the proportional coefficient matrix C. The parameters of the model are coupled with each other, and the differential evolution algorithm is used here for identification. In the case of a four-cylinder engine with a symmetrical exhaust manifold, the unknown parameters in the model are $K_1 \sim K_4$, $c_1 \sim c_4$, T_{lag} and d. It is possible to identify the UEGO parameters by a separate experiment so that the parameters to be identified can be further reduced to 8. The intake valve closing time of each cylinder is specified in the Simulink, and the A/F of the mixing gas is output after the engine model running in the GT Power. Four different random sequences were used as the intake valve closing time of the four cylinders so that the A/F of the mixing gas was evenly distributed within ±10% of the equivalent ratio. The A/F is sampled after the UEGO sensor at the confluence point of the exhaust manifold produces a stable periodic response. At the engine speed of 1600 r/min, 40 sets of stable input and output are collected as identification data for parameter identification.

DE is for dealing with the continuous optimization problem. By using real number coding, an initial population $\{X_{i,0} \mid i = 1, 2, \ldots, N\}$ is randomly sampled from the feasible solution space. Real-valued vector $X_{i,G} = (X_{i,G}(1), X_{i,G}(2), \ldots, X_{i,G}(D))$ represent an individual in a population at generation G. Where D is dimension of the objective problem, and N is the population size. Each individual in the population is a set of 8 parameters to be identified. The square sum of the error between the estimated output of A/F and the actual measured output is calculated as the first fitness index of the individual.

$$f(X) = 100 \times \sqrt{\frac{\sum_{i=1}^{N}(L_{sen,i} - \hat{L}_{sen,i}(X))^2}{N}} \quad (17)$$

For each individual X_i in the current population $\{X_{i,G} \mid i = 1,2,\ldots,N\}$, DE creates a mutant vector using "DE/rand/1" mutation strategies as shown in (18).

$$V_{i,G} = X_{r1,G} + F \cdot (X_{r2,G} - X_{r3,G}) \quad (18)$$

Where r1, r2 and r3 are distinct integers randomly selected from the range [1,N] and are also different form the index i. The scale factor F is a positive control parameter for scaling the difference vector.

After mutation, DE performs a binomial crossover operator on $X_{i,G}$ and $V_{i,G}$ to generate a trial vector $U_{i,G}$.

$$U_{i,G}(j) = \begin{cases} V_{i,G}(j), & R_j(0,1) \le P_{CR} \ or \ j = j_{rand}; \\ X_{i,G}(j), & \text{otherwise.} \end{cases} \quad (19)$$

Where j = 1, 2, ... , D, j_{rand} is a randomly chosen integer in [1,D], $R_j(0,1)$ is a uniformly distributed random number between 0 and 1 which is generated for each j, and $P_{CR} \in [0,1]$ is called the crossover control parameter.

DE selects the better one from the trial vector $\boldsymbol{U}_{i,G}$ and the target vector $\boldsymbol{X}_{i,G}$ to enter the next generation as shown in (20).

$$\boldsymbol{X}_{i,G+1} = \begin{cases} \boldsymbol{U}_{i,G}, & \text{if } f(\boldsymbol{U}_{i,G}) \le f(\boldsymbol{X}_{i,G}); \\ \boldsymbol{X}_{i,G}, & \text{otherwise.} \end{cases} \quad (20)$$

In order to test the accuracy of parameter identification with DE algorithm, the identification results are compared with those of the recursive least square (RLS) method. In RLS method, Different individual A/F L_{cyl} and the stable output of mixed A/F L_{sen} under the same conditions are taken as identification data for the parameter identification. Finally 4 sets of vector C were obtained corresponding to 4 cylinders' exhaust BDC. The comparison of the parameter identification results is shown in the Table II, where #1~#4 represent cylinder1~cylinder4 respectively.

Table 2. Parameter identification results

Parameters	DE	RLS			
		#1	#2	#3	#4
C_1	0.516	0.535	0.543	0.563	0.537
C_2	0.278	0.249	0.238	0.229	0.243
C_3	0.132	0.148	0.151	0.146	0.152
C_4	0.074	0.068	0.069	0.062	0.068

It can be seen from the identification result of RLS that the parameters of four cylinders basically satisfy the symmetry hypothesis. Because of the accuracy of the model and the interference in the identification, there are some errors in the mixed coefficient identification, but it can also reflect the influence of the exhaust of each cylinder on the mixed air-fuel ratio.

4 Estimation results of Kalman observer

The estimation results of the individual A/F are analyzed according to varied intake valve closing time as shown in Fig. 5. The intake valve opening time of each cylinder is 360 crankshaft angles, the cylinder fuel injection is constant, engine speed is 1600 r/min, the average output torque is 100 Nm, and the lift of intake valve is 8 mm. The IVCT of each cylinder changes as shown in Fig. 5(a). The mean zero white noise is added to the simulation to simulate the actual environment. As can be seen from Fig. 5(b) ~ (e), the large fluctuations of the A/F at the transition point increased individual cylinder A/F estimation error. After a short adjustment time, the error between the estimated value and the real value is stable within a certain range. The proposed Kalman Observer can suppress the system noise well, and since the maximum estimation error was less than 1% under steady-state conditions, the estimation performance was considered satisfactory.

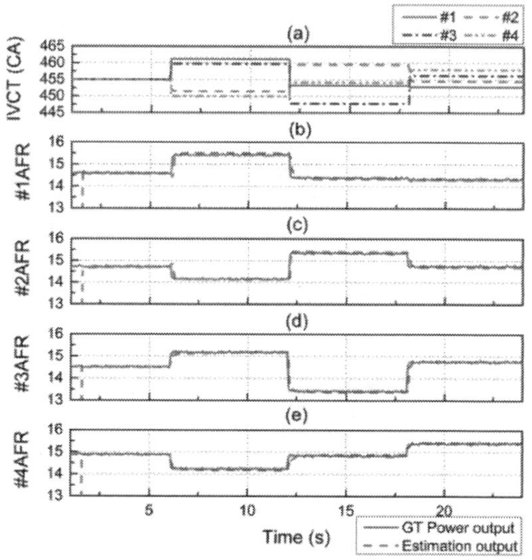

Figure 5. Individual cylinder A/F estimation results with respect to IVCT change (#1~#4 represent cylinder1~cylinder4 respectively).

5 Conclusion

In this paper, an individual A/F estimation approach for a four-cylinder gasoline engine with EMVT was developed for the maldistribution of the cylinder A/F. The main conclusions are as follows:

(1) The proposed Kalman filter observer allows estimation of the individual cylinder A/F from a single UEGO sensor at the confluence point of the exhaust manifold. The observer is also applicable to A/F estimates of other engines with symmetrical exhaust manifolds.

(2) The unknown parameters in the observer are identified by the differential evolution algorithm. Only a single UEGO sensor is used to complete the parameter identification, which simplifies the test and reduces the cost significantly compared with the common estimation methods.

(3) The proposed A/F observer can provide good estimation results under different valve opening duration, and has better ability to suppress the system noise, and the maximum error is less than 1% under steady-state condition.

Acknowledgment

This work was supported in part by the National Natural Science Foundation of China (grant number 51306090) and the Natural Science Foundation of Jiangsu Province, China (grant number BK20130762).

References

1. CW Wu, RH Chen, JY Pu and TH lin, "The influence of air-fuel ratio on engine performance and pollutant emission of an SI engine using ethanol-gasoline-blended fuels," *Atmospheric Environment*, **vol. 38**, pp. 7093-7100, (2004).
2. H Cho, K Min, SH Hwang and J Lee, "Prediction of the air-fuel ratio in transient conditions using a model of liquid fuel behaviour in the intake port of a spark-ignition engine," *Proceedings of the Institution of Mechanical Engineers Part D Journal of Automobile Engineering*, **vol. 214**, pp. 731-740, (2000).
3. L Benvenuti, M. D. D Benedetto, S. D Gennaro and A. Sangiovanni-Vincentelli, "Individual cylinder characteristic estimation for a spark injection engine," *Automatica*, **vol. 39** pp. 1157-1169, (2003).
4. J Kim, S Oh, K Lee, M Sunwoo, W Kim, C Lee and M Kim, "Individual Cylinder Air-Fuel Ratio Estimation Algorithm for Variable Valve Lift (VVL) Engines," *Sae. Technical. Papers.*, 2010-01-0785, (2010).
5. T Hara, T Shen, Y Mutoh and Y Liu, "Periodic Time-Varying Observer-Based Learning Control of A/F Ratio in Multi-cylinder IC Engines," *Springer International Publishing*, pp. 65-83, (2017).
6. W Schick, C Onder, L Guzzella, "Individual cylinder air–fuel ratio control using Fourier analysis," *IEEE Trans. Control Systems Technology*, **vol. 19**, no. 5, pp. 1204-1213, (2011).
7. R. Mallipeddi, P.N. Suganthan, Q.K. Pan and M.F. Tasgetiren, "Differential evolution algorithm with ensemble of parameters and mutation strategies," *Applied Soft Computing*, **vol. 11**, no. 2, pp. 1679-1696, (2011).
8. J Xu, S Chang, X Fan and A Fan, "Effects of electromagnetic intake valve train on gasoline engine intake charging," *Applied Thermal Engineering*, **vol. 96,** pp. 708-715, (2016).
9. X Fan, S Chang, L Liu and J Lu, "Realization and optimization of high compression ratio engine with electromagnetic valve train," *Applied Thermal Engineering*, **vol. 112**, pp. 371-377, (2016).
10. L Liu and S Chang, "Improvement of valve seating performance of engine's electromagnetic valvetrain," *Mechatronics*, **vol. 21**, pp. 1234-1238, (2011).
11. Y Geng and J Wang, "Adaptive estimation of multiple fading factors in Kalman filter for navigation applications," *Gps Solutions*, **vol. 12**, pp. 273-279, (2007).

AUTHOR INDEX